# BEHAVIORAL SYNTHESIS

## Digital System Design Using the Synopsys® Behavioral Compiler™

**David W. Knapp**

For book and bookstore information

http://www.prenhall.com

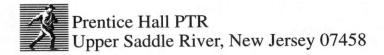

Prentice Hall PTR
Upper Saddle River, New Jersey 07458

Acquisitions editor: *Russ Hall*
Editorial assistant: *Maureen Diana*
Cover design director: *Jerry Votta*
Cover design: *Design Source*
Cover photo: *Dan McGanah courtesy of Direct Stock*
Manufacturing manager: *Alexis R. Heydt*

© 1996 by Prentice Hall PTR
Prentice-Hall, Inc.
A Simon & Schuster Company
Upper Saddle River, New Jersey 07458

The publisher offers discounts on this book when ordered in bulk quantities.
For more information, contact:

    Corporate Sales Department
    Prentice Hall PTR
    One Lake Street
    Upper Saddle River, NJ 07458
    Phone: 800-382-3419; FAX: 201- 236-7141
    E-mail: corpsales@prenhall.com

DesignWare, Design Compiler, Behavioral Compiler, FSM Compiler, HDL Compiler,
DesignWare Developer, Design Analyzer, COSSAP, dont_touch, and Synopsys
are trademarks or registered trademarks of Synopsys, Inc.
Express is a trademark of i-Logix Inc.
Verilog is a registered trademark of Cadence Design Systems, Inc.

Printed in the United States of America
10   9   8   7   6   5   4   3   2   1

ISBN 0-13-569252-0

Prentice-Hall International (UK) Limited, *London*
Prentice-Hall of Australia Pty. Limited, *Sydney*
Prentice-Hall Canada Inc., *Toronto*
Prentice-Hall Hispanoamericana, S.A., *Mexico*
Prentice-Hall of India Private Limited, *New Delhi*
Prentice-Hall of Japan, Inc., *Tokyo*
Simon & Schuster Asia Pte. Ltd., *Singapore*
Editora Prentice-Hall do Brasil, Ltda., *Rio de Janeiro*

# Contents

# Preface

Over the past three decades we have seen an astonishing increase in the scale of circuits we routinely design. This increase came about because of advances in lithography and silicon processing. The increase in the number of available transistors has resulted in numerous changes in design techniques and methodologies, as engineers have struggled to master the complexities of ever-increasing numbers of elements within a roughly constant development time.

During this period we have also witnessed the appearance of increasingly powerful synthesis and verification tools. These tools have helped us to master the possibilities inherent in our ability to manufacture ever-greater numbers of transistors. This is not always a comfortable process: engineers must choose between new, comparatively untried tools, which at least hold out the promise of handling the current level of complexity, and older, better-understood tools whose per-transistor effort level remains roughly constant.

This book describes methodologies for using a comparatively new class of synthesis tools, called variously 'behavioral synthesis', 'high-level synthesis' (HLS), and other less common names. These tools can radically reduce the level of effort needed to design a circuit of a given complexity; or conversely, they allow a much more complex design to be constructed with about the same effort. Behavioral synthesis does this by relieving the engineer of the burden of defining state machines and assigning operations to states. This allows the user to consider a variety of design alternatives, over a wide cost/performance range, with almost no effort beyond the construction of the initial functional description.

The increased abstraction of a high-level specification also radically decreases the size of the initial functional description, when compared to descriptions written at the next lower level of abstraction: usually by a factor of three to five, and sometimes much more. The result of this decrease in description size is a corresponding decrease in the number of bugs a user will tend to create; in the number of decisions that must be made; in the time it takes to construct the description; and in the amount of effort needed to understand it afterward.

The methodologies described here are based on hardware-description languages (HDLs). In order to get the most out of the tools, you need a 'good style' of writing

HDL descriptions. Good style requires an understanding of what's going on inside the behavioral synthesis system; it also requires that you understand how you will verify your design both before and after synthesis has been performed. These issues are closely interrelated. A useful analogy is to a discussion of good C language style, which would range all the way from how memory is treated to the design of algorithms for implementation on a machine with a single memory and a single instruction stream. Clearly, some styles of expression are better than others; but to understand why, we need at least some understanding of how our C program is going to be compiled, executed, tested, and debugged. The case is similar with HDL descriptions for HLS: we cannot expect to write good code by accident.

This book is intended for the working engineer, the engineering manager, and the student. The engineer will find in these pages ways to describe and synthesize designs using behavioral synthesis; in addition, he or she will find ways to characterize and debug design descriptions, and ways to get the most out of a behavioral synthesis tool. The engineering manager will obtain an introduction to a new way of thinking about digital design, a sense of the kinds of problems for which it is useful, and a premonitory glimpse of the kinds of difficulties that will have to be resolved. The student will be exposed to a description of a working, commercially viable tool, with emphasis on the nuts and bolts from an industrial and user perspective. All readers will get an introduction to a leading behavioral synthesis tool, the Synopsys Behavioral Compiler, which is used as a basis for discussion, and of which I was one of the builders. This book is loosely based on Version 3.4 of the tool; but in places I will refer to features that are expected to be present in future releases as well.

This is not an academic or theoretical treatment of behavioral synthesis. There are a number of very good general theoretical books and articles on the subject; [4], [5], [13], and [14] are good examples. This book is intended to provide a working understanding of an industrial approach to behavioral synthesis. Thus it is a supplement to a theoretical treatment rather than a substitute for one. By the same token, however, this book describes many real problems and details of behavioral synthesis that have received little attention in more theoretically oriented circles.

The first six chapters of this book contain a general description of behavioral synthesis and of Behavioral Compiler in particular.

**Chapter 1** is an introduction to the general design flow and the behavioral synthesis flow. This introduction will orient the reader and provide motivation and background material for the subsequent chapters.

**Chapter 2** discusses the inputs, structure and internal representations, and outputs of a behavioral synthesis system: the Synopsys Behavioral Compiler. This will help you to understand what is happening inside the synthesis process, and to understand the things that BC tells you.

**Chapter 3** gives a discussion of basic HDL description styles and constructs that accomplish specific design goals; for example, how to create a specific state diagram, how to cause a memory to be instantiated, how to handle I/O protocols, and so on. The emphasis here is on the ways in which HDL semantics imply various kinds of hardware structures.

**Chapter 4** discusses I/O timing in the HDL description and different models of equivalence between the pre- and post-synthesis designs. These interactions have important consequences for the ways in which HDL descriptions can be written and synthesized.

**Chapter 5** describes ways in which you can drive Behavioral Compiler by using commands and directives. This complements the discussion of Chapters 3 and 4, which show how to achieve specific goals using HDL semantics only.

**Chapter 6** gives a description of the kind of output the user of behavioral synthesis can expect to see. This chapter necessarily relies most heavily on the current Synopsys product, but the general strategy and descriptions of basic reporting styles should prove useful beyond that context as well.

The latter part of this book is divided into case studies; each has been chosen for its ability to illustrate some important feature of behavioral synthesis or of the overall process. The case studies presented here are also included on the diskette that accompanies this book; you can study them, simulate them, synthesize them, and change them as you will. The examples used are the following.

**Chapter 7** describes an FIR filter, which illustrates a basic behavioral description style for the superstate-fixed I/O mode, which is the simplest of the I/O modes Behavioral Compiler supports. It also describes some common pitfalls, and some basic performance-tuning techniques.

**Chapter 8** describes an IIR filter, which illustrates basic behavioral specification style for cycle-fixed I/O mode and the construction of a handshaking I/O protocol. More advanced performance-tuning techniques are also used, including the use of pipelined components.

**Chapter 9** describes an inverse discrete cosine transform. This chapter begins with a small program fragment (a pair of matrix multiplications) written in C, and then gradually transforms it from a naive translation that doesn't work very well into a much more sophisticated version. This example also provides you with experience in using memories.

**Chapter 10** describes the Data Encryption Standard, which illustrates the synthesis of a design consisting almost entirely of random logic. This example also shows the use of DesignWare to encapsulate random logic for sharing.

**Chapter 11** describes a packet router, which is an example of a control-dominated design with memory.

There are also two appendices. Appendix A gives a brief description of the use of DesignWare in conjunction with the other Synopsys tools; and Appendix B gives an overview of the synthesizable subsets of VHDL and Verilog that are supported by the current Synopsys products.

## Typographical conventions

It is usually pretty obvious when a word or other symbol is coming from a different language or namespace than that of the base text. For example, if I use the term 'entity' in the context of a VHDL hardware description, most readers will not be fooled: they won't go looking for the definition of 'entity' in a dictionary. Instead, they will semi-automatically refer to the namespace of VHDL constructs. This is just common sense and shared background.

I have therefore relied on the following guideline: where the namespace of a symbol is ambiguous, and there is a sense under which the ambiguity might be resolved incorrectly, I will use a different font to indicate that the symbol is taken from another namespace than that of written English. For example, I might discuss the architecture of a CPU; here there is an ambiguity, because I might mean either the VHDL `architecture`, which is taken from VHDL's namespace, or else the instruction set's treatment of registers, status, and so on. In a case like that one, I will use a special font (`courier`) to denote the VHDL construct. Sometimes it might not matter: a VHDL `loop` is in most cases just another loop, and deserves no special treatment. In such a case I will use the convention that makes the text flow most smoothly.

The other case where I will use typography to set off a symbol taken from a non-English namespace is where a sentence might become awkward or grammatically incorrect if the symbol is not set off in some way. For example, in this sentence the mathematical variable $a$ would look very strange if it were not set in italics, and the VHDL variable `index` would be misleading without the special font.

But excessive use of typographical variations blights many an otherwise readable text. Setting off every keyword of such a populous language as VHDL would quickly result in the textual analog of an 'angry fruit salad' graphical user interface. So where neither grammar nor ambiguity constrains the choice of language and typography, I will stick to standard English set in plain Roman.

## Acknowledgements

A book like this one is necessarily the product of a great deal of consultation, review, and discussion. I can only begin to acknowledge the contributions of the many reviewers of the various iterations of the manuscript: most prominent among them were Giovanni De Micheli, Wayne Wolf, David Black, Marc Barberis, Mike Craven, Tony Dimalanta, Pradeep Fernandes, Mick Robinson, Margaret Marek-Sadowska, and the members of my immediate team.

A special acknowledgement must go to Kristen McNall, who diligently and with great perseverance helped with debugging the text, and also went to heroic lengths to ensure that the examples were correct and consistent in two languages. I would also like to acknowledge the contribution of Marc Barberis, to whom I am indebted for virtually all of my understanding of the digital signal processing domain; If I made mistakes in that domain they persist in spite of Marc's best efforts. Pradeep Fernandes, Hazem Almusa, and Tony Dimalanta helped me to understand and finally to write about many of the details of BC that I did not understand myself. Ronald Niederhagen and Srinivas Raghvendra both helped me with some highly esoteric issues.

I would like also to acknowledge the contributions of the BC development team: Tai Ly, Ron Miller, Don MacMillen, Reiner Genevriere, Chaeryung Park, Taewhan Kim, Peter Zepter, and Shu-Mei Cheng. Without the contributions of these individuals there would be no BC to write about. Finally, I want to thank T.J. Boer for his support: there would be no book at all but for his efforts.

# Chapter 1

# Introduction

This chapter contains an overview of digital circuit synthesis, which is the context within which a behavioral compilation system operates. This context is described concretely, in terms of a set of tasks that must be performed by an 'ordinary' designer. Where there are differences between my 'ordinary' flow and your own, please bear with me, fill in the blanks, and mentally correct the differences: no concise description can convey everybody's actual experience. Each of the tasks will be described in general terms. Then we will turn to a discussion of the role of verification and validation, and give a brief overview of the tools available for verification and validation, and the ways in which they can be used. This will provide you with a context in which to place the subsequent discussion.

After the overview we will zoom in on the design flow in the near region of behavioral compilation; the general structure of the behavioral compilation task will be described, with emphasis on the relationship of the behavioral compilation task to its neighbors.

## 1.1 Design Flow

The main synthesis tools and the data flows between them are shown in Fig. 1.1. The data flows are depicted here as one-way; in fact, user-mediated iteration may occur at and across all of these levels.

Under normal circumstances, the rough outlines of the design should be explored at the highest possible level, and then increasingly fine-grained optimizations should be applied at increasingly lower levels. Experience indicates that a wider variety of designs can be explored by iterating at the higher levels (e.g. when functionality is partitioned into blocks or processes) than at the lower levels (e.g. at logic synthesis time), where the high-level decisions have already been made. Of course it may be necessary at any stage to go back to some previous stage; we try to avoid this, and the figure doesn't show it, but it would be unrealistic to pretend it never happens.

At the top of Fig. 1.1 are two examples of *domain-specific* synthesis tools. The ones shown here, dataflow oriented synthesis and control flow oriented synthesis, are those most likely to be familiar to the reader.

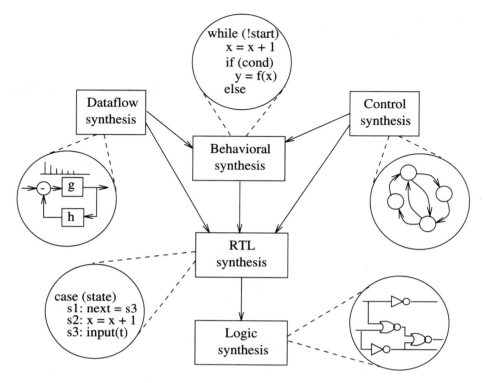

**Figure 1.1.** Relationships between synthesis packages

In a *dataflow oriented* synthesis package, the input represents data as streams of samples, as opposed to single samples; and operations on data are stream-oriented operations. So, for example, an expression like $x = axz^{-1} + bu$ is interpreted as meaning that two synchronized streams of samples $axz^{-1}$ and $bu$ are fed into an adder, which then produces a third stream of samples $x$, each sample of $x$ being the product of the corresponding samples of $axz^{-1}$ and $bu$.

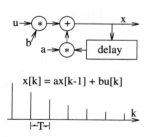

An example of a dataflow oriented language is Silage [22]; and an example of a dataflow oriented graphical capture tool is the COSSAP Block Diagram Editor [8], [10], [20]. The underlying idea of a dataflow synthesis tool is that data streams are transformed into other data streams in a steady flow of data through the system being synthesized; within limits, one can say that the same operations are applied to all input samples. This kind of representation is useful, for example, in describing signal processing applications such as filters and modulators.

The abstraction of signals into streams enables very fast simulation. The importance of this becomes apparent when we consider the range of time scales over which stream-oriented simulation can be applied. For example, an engineer might want to model an equalizer, which filters a stream with a frequency of many megahertz through a model of a time-varying communication channel whose properties change on a time scale of seconds. Thus in order to model the response of the system to channel variations, the behavior of the system must potentially be modeled over many millions of samples.

Synthesis of dataflow oriented descriptions consists of mapping a stream oriented input description into a time-oriented description, which can then be synthesized either at the behavioral level or the register-transfer level. Typically the output of dataflow oriented synthesis is an HDL text that can be synthesized by either a register-transfer level (RTL) or a behavioral synthesis tool.

In a *control flow* oriented synthesis package the emphasis is on states and transitions between states. Input descriptions for control flow oriented synthesis can be either language based, as in [2]; or graphical, as in [7].

The states of a control flow oriented description may be hierarchical. At the upper levels of abstraction the states are abstract states of the application domain: for example an aircraft control system might have a state called 'landing approach', with transitions to 'go around' and 'rolling'. These states can then be decomposed into further hierarchical states. At the lowest levels the states may or may not represent individual states of an underlying FSM circuit.

Typically the output of control flow oriented synthesis tools is an HDL text that can be passed to either RTL or behavioral synthesis. If the states at the lowest level of the hierarchy directly correspond to circuit states, then the output of control oriented synthesis can be sent to register-transfer synthesis; otherwise, behavioral synthesis must be done.

The main difference between a control oriented synthesis system and a dataflow oriented system is that the dataflow oriented system regards data as a collection of streams, where each element of each stream is to be processed in much the same way as the other elements of that stream; whereas in a control oriented system the data and the treatment of the data tends to be much less regular, with emphasis on different behaviors in different states.

*Behavioral synthesis* [1], [21], [22], [13], [23], [16], [14] can be used as a back end for both control oriented and dataflow oriented synthesis, or as a general-purpose tool in its own right. Typically the input to behavioral synthesis is language-based. VHDL and Verilog are the most widely known input formalisms for behavioral synthesis.

```
x = 0;
forever begin
  u = inp; // input
  x = a * x + b * u;
  outp <= x; // output
  @(posedge clk);
end
```

If behavioral synthesis is used as a back end for another type of tool, it is usually advisable to make the interchange format an HDL; there are many benefits associated with standardized, simulatable, and readable interchange formats.

Other languages that have been used are variations on APL, Pascal, C, ISP, and a number of less well-known languages. The primary advantages of using languages like Verilog and VHDL are that there are high-quality simulators available and in common use, there are existing register-transfer level and logic synthesis tools that accept these languages as input, and there is an existing customer base of designers, who are familiar with the use of these languages.

The core activity of behavioral synthesis is *scheduling*, in which operations are assigned to states. This is by contrast to register-transfer level (RTL) synthesis, in which operations have already been mapped to states by the user. Scheduling can create new states as well; whether it does so depends on the user's constraints.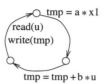

The states being discussed in the behavioral and RTL synthesis contexts are actual machine states, i.e. their transitions are on the edges of the underlying physical machine's clock, as opposed to some other model of time.

For example, in a system that does audio processing, the audio sample stream might have a rate of 45 kHz, which is the rate of the stream of input samples. In the implemented system there might be a filter employing many arithmetic, logical, and memory operations; each of these operations might take a single cycle of a much faster machine clock, perhaps running at many megahertz, which strobes the registers and memories that store the intermediate data of the filter. Thus there are two clocks in consideration: the sample clock and the machine clock.

Behavioral and RTL synthesis consider the machine clock, while synthesis of stream-oriented designs is based on streams of samples. Control-dominated synthesis may consider either a derived clock or a machine clock.

Behavioral synthesis also performs *allocation*, in which operations are assigned to functional hardware and data values are assigned to storage elements. Typically register and storage allocation is only done by behavioral synthesis packages. Operator allocation can be done at either the RTL or at the behavioral level, but the range of options open to behavioral-level operation allocation is greater than it is at the RTL. For example, a behavioral package can trade off latency for area, which RTL synthesis packages cannot do, and trade off register area for combinational and interconnect area, which is theoretically possible at the RTL but which in practice has severe limitations.

In RTL synthesis the state graph is manually defined and operations are tightly bound to particular states or state transitions; data values are also tightly bound to particular storage resources. RTL synthesis can serve as a back end to behavioral, control flow, or dataflow synthesis; it can also be used as a front end in its own right. RTL level descriptions give the user a great deal of control, but they are tedious, cumbersome, verbose, and prone to error.

Following RTL synthesis is combinational logic synthesis. Here a combinational logic function is optimized for propagation delay and area. So logic synthesis forms the back end for RTL synthesis, which is a back end for behavioral synthesis, which in turn can be used as a back end for control and data oriented synthesis.

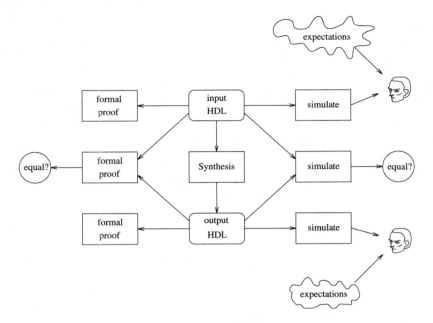

**Figure 1.2.** Relationships between synthesis and verification

## 1.2  Simulation and Verification

Even when our designs are being produced by software, we still like to be sure that they do what we want. There are many types of tools that help us to verify and validate our circuits; typically in the synthesis context we speak of two types of verification tools. First, *simulation* tools are used to test the response of the design to particular inputs. This is by contrast to *formal verification* tools, which test mathematically defined properties, such as the Boolean function computed by a circuit or its equivalence to some 'golden' model.

Simulation is the most practical way to validate the user's input against expectations; for this reason any primary input formalism has to have a simulator. This is shown at the upper right side of Fig. 1.2. In the dataflow domain such a simulator is the COSSAP Stream Driven Simulator in the control flow domain the Express analyzer is the best known example. For users whose primary input is at the behavioral level and below, there are many available VHDL and Verilog simulators. Simulation has an inherent coverage problem: specific inputs and states are simulated, so in ordinary-sized designs there will be input/state combinations that simulation does not cover.

Formal verification has the advantage that it proves the properties of a design without reference to specific input vectors. However, it is not as useful in validating the behavior of primary input descriptions. For example, one might prove that a control oriented design was deadlock free, or that it performed the specified function. Even so, the design might not do exactly what you want: there may also be bugs in the your ideas. Formal verification tools are usually divided into two categories: those that check designs, and those that check implementations. A formal *design verification* tool checks for the presence or absence of specific properties. For example, such a tool might check a control-oriented design to prove that it could never enter an undefined or deadlocked state. A formal *implementation verification* tool checks for consistency between an initial representation and a final representation. For example, such a tool might compare the Boolean function of an unoptimized logic network with the Boolean function of the same network after optimization; the idea would be to prove that the output of the second network would always be the same as that of the first for all defined inputs.

The most widely used formal verification tools do combinational comparison. Here the property being proved is equivalence of two implementations; one, e.g. an input description before optimization, and the other an automatically or manually optimized version. This is shown at the center left of Fig. 1.2. The main drawback of formal verification tools of this kind tends to be run time and memory requirements, which are often excessive.

The presence of state information makes formal verification above the level of combinational logic much more difficult. In effect, a VHDL **process** or Verilog **always** block is as hard to prove correct as an equivalent program written in C or Pascal; with the added complexity that C and Pascal are single-process languages, where VHDL and Verilog are multiprocess languages. Hence the best way to validate behavioral synthesis today is simulation (center right, Fig. 1.2).

## 1.3   RTL and Behavioral Design

Until the advent of commercial behavioral synthesis tools the typical design flow had something of a gap between the level of domain-specific synthesis tools and HDL-based RTL synthesis tools. Behavioral synthesis can be thought of as bridging that gap. It can also be thought of as providing a higher level of abstraction for the designer's input to logic synthesis; this is an appropriate way to think about the design flow in areas where domain-specific tools are not applicable or available. Here the input to synthesis is in the form of a hardware description language (HDL) text. We will now turn to the HDL-based design flow and describe it in more detail, with emphasis on the tasks and ways in which behavioral synthesis can assist the user.

Let us suppose that we begin with an initial model written in a general-purpose programming language such as C or C++. The purpose of the initial model is to define and test the functional aspects of the design's behavior. For example, if we are designing a filter, we could use the initial model to ask questions about bit widths, operation orderings, and rounding strategies; if we were designing an ATM packet router, we might

```
x = 0;
while (!reset) {
    u = read(input);
    x = a * x + b * u;
    write(output, x);
}
```

ask questions about the number of operations necessary to unpack a routing field and store a packet in the appropriate queue.

Timing, state, and other important properties are modeled in abstract ways; for example, a queue may be modeled as a block of memory, without detailed bit representations, size limits, or access conflict models.

Once the we are satisfied that the initial model does the right thing to within the limits of its abstractions, it can be translated into a hardware-description language. Then we can model concurrency and time in a natural and accurate way. The first task of the HDL model is to get interfaces between modules and module timings right, using a simulator.

An HDL description can also be fed directly into a synthesis tool. This requires the use of a *synthesizable subset* of the HDL: many constructs of the HDL are useful for simulation and debugging, but are not synthesizable. Examples of unsynthesizable constructs are assertions, file I/O, and simulator display functions; propagation delays; some data types such as continuous values, and attributes such as signal stability and some event attributes. Appendix B gives an overview of the synthesizable subsets of VHDL and Verilog.

An input description at the behavioral or register-transfer level usually consists of a number of processes (in Verilog, these are **always** blocks). Each process is synthesized as a separate finite-state machine (FSM) or as a combinational block, depending on whether it makes use of state information. The processes may be connected together by means of nets and random logic. Logic outside processes is synthesized using gate-level techniques.

Consider a process that describes a finite-state machine. It will contain an implicit or explicit representation of what its states are, and what its responses to various inputs will be in each of its states; these responses will be its outputs and next states. In a behavioral description, the states are implicit, and the number of states can be varied by an optimization algorithm; additionally, the operations are not rigidly mapped to states, and optimization algorithms can move operations from one state to another.

In behavioral synthesis the state diagram and assignment of operations to states are both constructed automatically. The big distinction between behavioral and RTL synthesis is that in RTL synthesis the user must decide what the states are, what the state transition graph is, and what operations happen in each state.

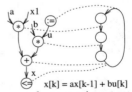

x[k] = ax[k-1] + bu[k]

In RTL synthesis, the user must also decide how registers and memories are to be used. Specifying states, transitions, operation-to-state mappings, and variable usage is all obligatory at the RTL but automatic in behavioral synthesis. There are important tradeoffs here. For example, a simple filter might require two multiplications. If you allowed two cycles you could use a single multiplier; but if you allowed one cycle you would need two multipliers. Here are two HDL descriptions of a filter $x[k] = ax[k-1] + bu[k]$; on the left, the RTL, and on the right the behavioral HDL.

```
always begin                          always begin
    case(state)                           x = 8'b0;
        2'b00:begin                       while (stop == 1'b0) begin
                x = 8'b0;                     u = inport;
                if (stop != 1'b0)             x = a * x + b * u;
                    next = 2'b01;             outport <= x;
                else                          @(posedge clock);
                    next = 2'b00;         end
            end                       end
        2'b01:begin
                u = inport;
                x = x * a;
                next = 2'b10;
            end
        2'b10:begin
                utmp = u * b;
                next = 2'b11;
            end
        2'b11:begin
                x = x + utmp;
                outport <= x;
                if (stop != 1'b0)
                    next = 2'b01;
                else
                    next = 2'b00;
            end
    endcase
    state = next;
    @(posedge clock);
end
```

As you can see, there are many more opportunities to make mistakes in the RTL. It is longer; the flow of control is less clear; the operations have to be carefully distributed among the states; and you have to manage the next-state function yourself. In a more complicated example the difference is even more pronounced.

The difficulties associated with constructing good RTL descriptions are often exacerbated by additional considerations, such as conditional operations, pipelining of components and/or loops, multicycle and sequential operations, memory operations, different state transition graph paths being taken under different conditions, the presence of memories, and interface timing constraints.

After the states and the state transition graph have been established, operations of the design can be *scheduled*, i.e. assigned to states. Operations that can be scheduled include arithmetic and logical operations, register reads and writes, memory reads and writes, I/O reads and writes, loop boundaries (loop begin, end, exit, and continue 'operations'), and function and procedure calls.

Operations must be scheduled so that nothing happens out of order and so that overall performance and cost are optimized. In RTL synthesis all of this has to be done manually, by ordering the statements in the source HDL. In behavioral synthesis most of the non-I/O operations are scheduled automatically. I/O operations are more constrained, because they define interface protocols; they must be scheduled in much more structured and restricted ways than other operations.

Once operations are scheduled they can be mapped onto shared hardware. This is done automatically in both RTL and behavioral synthesis. Sharing saves gates and can under some circumstances decrease propagation delay as well, as e.g. when multiplexers can be eliminated.

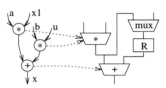

In RTL synthesis register usage is also determined by the user. A register is inferred from the RTL if in any state a variable could be read before it is written; so you have to be very careful with variable references and assignments, and figure out all of the register sharing yourself. This means keeping a clear idea of what data is in which variables at all times, and reusing variables when they become 'dead'. This can get very complicated in the context of a large or complex state transition graph. Furthermore, it is strongly affected by the schedule of operations, and the assignment of data to registers also affects the costs of moving data around, and hence of operation sharing.

By contrast, in behavioral synthesis the variable lifetimes and the minimum set of registers needed to store the variables are computed automatically; variables are also assigned to registers by an optimization algorithm.

After operations and data values have been shared, the netlist for the design can be constructed and particular implementations selected for the components. For example, an adder might be implemented using ripple carry or carry lookahead; this is usually a timing-driven decision, but if the timing permits a cost-driven decision can be made as well. This can be overridden manually, but the software does a pretty good job: manual intervention at this stage is not common.

So the main differences between behavioral and RTL design are that the states, state transition graph, operation scheduling, and register sharing are all constructed automatically in behavioral synthesis and manually in RTL synthesis. These lead to a big difference in the amount of time it takes to construct a good HDL specification, because so many of the details (such as where state boundaries occur) are automatically optimized.

The differences between behavioral and RTL descriptions also have a profound effect on the readability of the HDL description, as we saw above. Behavioral descriptions tend to be one-third to one-fifth as long as equivalent RTL descriptions, the flow of control is clearer, and operations can be placed for readability instead of for efficient mappings to time. This results in big economies when the design must be debugged or modified, because it can be more readily understood.

When scheduling is automatic, changes of such important parameters as the number of states in a pipeline can be done by changing a single constraint, rather than by picking apart and then reassembling a complex HDL description of a state graph. Thus you end up with increased confidence in your design, a shorter debugging cycle, potentially higher quality (you can do more exploration in the same time and are less likely to make mistakes), and a cleaner design.

The netlist is then *compiled*[1], i.e. optimized at the gate level. There is no essential difference between gate-level optimization of a design that began at the RTL and one that began at the behavioral or some higher level. They are just networks of gates, which must be optimized.

Logic-level optimization is done using techniques of Boolean minimization followed by technology mapping. The design might then be *retimed* if the user needs to meet a tight timing constraint. Logic optimization and retiming can be done more than once, with hierarchies being grouped and ungrouped in between, and constraints changed; this is an iterative process that can potentially be repeated many times.

Following logic optimization, the design may be modified to increase its testability; it may be written out for simulation at the gate level or timing analysis; or it may be written out for layout.

## 1.4   Behavioral Compiler Design Flow

Fig. 1.3 shows the overall design flow for the Synopsys Behavioral Compiler schematically. We defer consideration of most of the details to the next chapter: here the purpose is to give you a specific and concrete outline, a road map if you will, of how behavioral synthesis is done in Behavioral Compiler.

We begin with an input description written in VHDL or Verilog. The input description can be the result of processing a higher-level description written in a domain-specific language, or it can be a primary description, i.e. constructed from scratch by the user. We will assume that the user has already simulated the HDL description and is satisfied with its behavior.

First, the HDL text is *analyzed*. This step does the parsing and some of the semantic analysis. A symbol table is built; syntactic errors, type clashes, undefined functions, etc. are reported at this time.

The design is then *elaborated* for scheduling. Elaboration for scheduling is different in detail from elaboration for simulation or for RTL synthesis; the result of elaboration for scheduling is a mixed control/dataflow representation of the design. This representation makes control flow and potential data parallelism explicit.

The design is then *constrained* by the user. Explicit user constraints supplement constraints that are implicit in the HDL text; for example, the HDL text might show one I/O operation preceding another, while an explicit user constraint might state that the first should precede the second by exactly five cycles. If the HDL text was automatically generated, e.g. by a dataflow-oriented synthesis tool, the constraints may be automatically generated as well.

---

[1] In the Synopsys terminology to `compile` means to perform logic optimization.

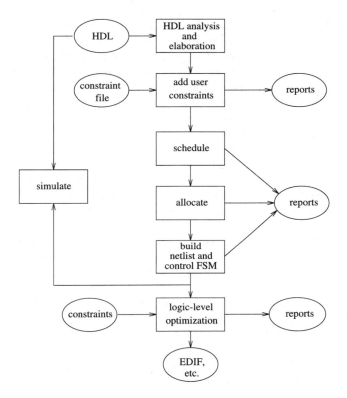

**Figure 1.3.** Behavioral Compiler: flow

In the next step, a timing model of the dataflow portion of the circuit is constructed. The dataflow is transformed into a gate-level netlist and mapped to the user's chosen target technology; then a timing analysis tool is run. The user can see the results as a table of propagation delays.

This early analysis of the timing of mapped hardware allows BC to schedule chains of operations within the clock cycle the user specifies. This flow is called *technology-specific* scheduling, to differentiate it from *technology-independent* scheduling; it is one big advantage BC has over its competitors.

The design can also be manually constrained after it is timed: perhaps the extra information will lead the user to constrain it after timing rather than before, or to modify constraints specified before timing analysis has been done.

The design is then *scheduled*. In this step states are constructed and the assignment of operations to states is performed. The user can choose to allow the scheduler a variable amount of freedom to change the I/O timing of the design; this choice must be made at the time scheduling is invoked.

Next, the design is *allocated*. In this step modules and registers are constructed, and mappings from the logical operations and data to the netlist modules and registers are defined. The output of allocation is a logical description of a datapath that consists of abstract resources and data transfers. Don't be confused: many writers distinguish between the allocation of resources and the binding of resources to operations and data values. The usage within the Synopsys team has been that the word 'allocation' covers both activities.

The netlist for the design is then constructed, and the con-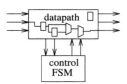
trol FSM is built. Depending on the pathway chosen for the
FSM synthesis, it is either a gate-level network or a state table when this is finished; it is then connected to the allocated
and interconnected hardware. The result is a design in which
every behavioral process has been mapped to a datapath and
a control FSM.

*Reports* on the schedule, allocation, and state graph can then be generated and examined by the user. These reports give estimates of the cost and speed of the design, and describe the schedule, allocation, and state graph in detail. If the reports are satisfactory, the user can go on to gate-level optimization. But gate-level optimization usually takes a lot more time than scheduling, so it is common to do some exploration at this stage, iteratively reconstraining the design and rescheduling it with the new constraints. It is also possible that changes to the HDL are desirable: in which case the cycle would begin again at HDL analysis.

The netlist can also be viewed at this stage, using e.g. the Synopsys Design Analyzer. This will help the user to understand what is happening in the design during simulation and gate-level synthesis; it may also suggest improvements that could be made in the HDL input and constraint scripts.

Another thing that can be done at this stage is RTL simulation. This is important for two reasons. First, most users want to be certain that the synthesized design still passes functional acceptance criteria. For example, a process's I/O timing may have been transformed by scheduling; that case, you want to be sure that the transformation doesn't break the I/O protocols through which the process interacts with its environment.

Second, RTL simulation is much faster than gate-level simulation, so it is preferable to find out about any problems now rather than after logic optimization. RTL simulation is done using HDL text generated after scheduling and allocation, but before logic optimization. The timing models simulated at this level are cycle-based; they use simple zero-delay timing models and cannot tell you, for example, about setup or hold timing violations.

The next step is logic optimization. This begins with FSM state assignment for the control unit; then implementations are selected for complex operators; then Boolean optimization is done; and finally the logic is mapped to the target tech-nology. At this time it is possible to do timing-accurate simulation, static timing analysis, testability insertion, test generation, layout, and other downstream tasks.

## Summary

At this point you should know how behavioral synthesis fits into the overall design flow; what scheduling and allocation do for you; the series of steps by which Behavioral Compiler will process your input; and of some of the features that a good behavioral synthesis system should have.

# Chapter 2

# Behavioral Compiler

In this chapter the Synopsys Behavioral Compiler (henceforth, BC) is described in terms of its inputs and outputs, its capabilities, and its internal structure. This exposition will help you by giving you a clear mental model of how BC works. This model will be useful in the later chapters, because you will have a conceptual framework upon which you can hang the details; it will help you to understand what is going on when your design is processed; it will help you to understand what error messages mean, it will help you to debug your design when something goes wrong; and it will help you to design good inputs to BC. This chapter does not provide an exhaustive description: it provides an overview and a framework. Succeeding chapters will provide deeper coverage of specific topics.

Strictly speaking, BC is a collection of functions embedded in a program henceforth called 'bc_shell', which serves as a text-driven user interface for HDL processing, scheduling, and compilation. You can also use the Design Analyzer (DA) tool, which provides a graphical user interface for the same functions. There is no internal difference between the capabilities of the two user interfaces, except that you can display circuit diagrams using DA. From here on we will use the term 'bc_shell' to denote both of these, with the understanding that the same textual commands can be typed into both, either interactively or by using scripts.

## 2.1 Inputs

The flow of data through BC is shown in Fig. 2.1, where bc_shell is represented by the rectangular boxes, and data files, etc. are represented by ovals.

BC gives you three basic classes of input mechanisms. These are the HDL text itself, the bc_shell command language, and comments (pragma directives) embedded in the HDL text.

The input HDLs supported by BC are the synthesizable subsets of VHDL and Verilog. An input description for BC should consist of a VHDL entity-architecture pair or a Verilog `module`. The architecture or module must contain at least one process (Verilog `always` block) for BC to be able to do anything with it: BC operates on processes. The input may contain any number of processes, as well as logic external to the processes. BC does not consider any interactions between

14

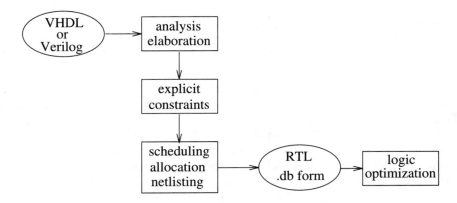

**Figure 2.1.** Flow of information through BC

processes. Because of this BC can operate on one process at a time, without ordering problems. Netlist constructs outside processes will be instantiated, and BC will do nothing further with them until logic optimization. Henceforth we will consider only one process, which we will call 'the BC process' where necessary.

Analysis of the input HDL is done using the bc_shell command **analyze**, in the top center box of Fig. 2.1. BC reads in the HDL text, parses it, and informs you of any syntax errors. It can also inform you of some semantic errors; for example, type clashes or undeclared variables. The **-f** flag specifies which HDL you are using.

```
bc_shell> analyze -f verilog mydesign.v
```

The design is then *elaborated*. This step builds an annotated circuit description. In BC, elaboration has two modes. The first mode is RTL elaboration, which results in a circuit that can be passed directly to Design Compiler for logic optimization. The second mode of BC elaboration is for scheduling (**-s** flag), which results in a description that can be scheduled.

```
bc_shell> elaborate -s mydesign
```

The **-s** flag can be overridden for a process by an attribute **rtl** attached to the process. In VHDL, this attribute is attached directly; in Verilog, an **always** block can be marked as RTL with a pragma.

```
/* synopsys resource R : synthesis_type = "rtl"; */
```

Declaring a process to be RTL switches elaboration of the process to RTL mode. This allows the user to mix behavioral and RTL processes in a single design; the **-s** flag is used to perform both kinds of elaboration with process-by-process determination of elaboration mode.

The user may then constrain the design. First, a clock net must be declared and its period stated. This is not optional; it must be done before scheduling is invoked.

```
bc_shell> create_clock clk -period 9
```

Second, operations can be forced to fall into specific control steps (for now, think of control steps as corresponding to clock cycles, or states. They are defined more precisely in Section 2.2.1), and pairs of operations can be constrained with respect to one another.

```
bc_shell> set_cycles 3 -from op1 -to op2
```

Third, implementations of components, e.g. of adders and subtracters, can be forbidden. This can be used to prevent the use of the component or force the use of an alternative when timing analysis is done: which is described below.

Fourth, scheduling strategy options are set. Finally, wire loading models can be set to reflect the expected wire loadings of the finished design.

At some point between elaboration and scheduling, the design's timing properties are analyzed. This step is needed so that the scheduler can tell whether specific chains of cascaded operations can be connected to one another combinationally, i.e. without registers between the operations; such *chaining* cannot be done if the summed delay along the proposed combinational path is longer than the clock cycle.

Timing analysis is based on mapped logic: BC is a technology-specific scheduler. Thus BC has accurate information to drive its operation chaining decisions. The effects of this can clearly be seen at logic synthesis time and in the short latencies that BC can achieve.

The timing analysis step is not shown on Fig. 2.1, because it can be invoked manually by the user at any time after elaboration and before scheduling; alternatively, the user can let it be invoked automatically by the scheduler.

```
bc_shell> bc_time_design
```

BC uses *bit-level* timing models in its timing analysis. This is in contrast with *lumped* or word-level timing models, which represent operation delays as a single number. The level of accuracy obtained by bit-level modeling is significantly better than with lumped models, particularly where there is random logic present and where there are chains of operations.

After timing analysis BC will list the combinational chains and their timing. This report can be useful in changing the clock cycle (if the clock cycle happens to be a free parameter), in changing the technology family (if technology is a free parameter), in estimating the required number of cycles to complete a particular piece of the computation, and as a guide for manual implementation selection.

Timing the design also tells the user if there are operations whose combinational delay exceeds the clock period. Such *multicycle* operations incur both area and timing penalties. To avoid these penalties, the user can either change the operation being performed, the implementation of the component, or the clock cycle.

The design can then be scheduled. In this step, operations are mapped to control steps; two operations mapped to different control steps can share a single multiplexed hardware resource. Sharing reduces cost; the primary goal of scheduling is to construct a schedule that allows for a minimum-cost (highly shared) implementation that still meets performance requirements.

We will now turn to the process by which scheduling is actually done. After that we will return to the user's view of the process, and discuss what outputs you can expect to see.

```
bc_shell> schedule -effort low
```

## 2.2  Behavioral Compilation: Internals

In the previous section we briefly sketched inputs for behavioral synthesis. We now turn to a discussion of the internals of behavioral compilation. This will establish a terminological base for further discussion, and also provide an intellectual basis that will help you understand what happens during behavioral synthesis. You need to know this, because it will help you understand what you are doing and why.

### 2.2.1  Representations

In scheduling, the operations of the original design are mapped onto control steps. A *control step* (abbreviated to *cstep*) is an integer representing an abstract machine state or clock cycle; two operations mapped to the same control step are concurrent. This is shown in Fig. 2.2, where four arithmetic operations are all concurrent in the first cstep.

The solid horizontal lines of Fig. 2.2 and Fig. 2.3 represent cstep boundaries. The csteps themselves are given numbers $0, ...i, i+1, ....$ Thus operation *mul2* of Fig. 2.2 falls into the cycle following operations *add2* and *sub2*.

The mapping of control steps to states is approximately one-to-one. There are important exceptions to this rule; but for now, you won't get into too much trouble by equating control steps, clock cycles, and states. You should remain aware that in section 2.2.5 this simplifying approximation will be retracted.

The term *latency* is used in the context of scheduling to denote the number of csteps, or sometimes states, falling between two events or operations.

The *control-dataflow graph*, henceforth CDFG, is used by BC as an abstract representation of the desired circuit behavior. It is supposed to do so without bias,

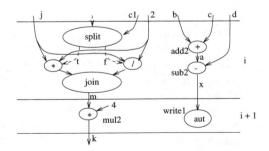

**Figure 2.2.** Simple CDFG example

so that any feasible schedule can be constructed by BC. In the source HDL and in RTL internal representations, by contrast, operations are given a 'schedule' just by their positions in the HDL text. These positions may be largely accidental. For example, consider this HDL text, which corresponds to Fig. 2.2.

```
Verilog                     VHDL
if (c1) m = j * 2;          if c1 then m := j * 2;
else m = j / 2;             else m := j / 2; end if;
a = b + c;                  a := b + c;
x = a - d;                  x := a - d;
@(posedge clock);           wait until clock'event
                                    and clock = '1';
m = j * 4;                  m := j * 4;
aut <= x;                   aut <= x;
```

In the HDL text, the addition and subtraction operations precede the clock edge, which is followed by an output write to the port `aut`. If we were to simulate it as it stands, the addition and subtraction would occur in the same cycle. But this schedule is an artifact of the way the HDL text is written, because the external (black box) behavior of the circuit would be indistinguishable from one in which the subtraction happened after the clock edge.

If we were to implement this HDL description using RTL techniques, i.e. without rescheduling the operations, this artifactual schedule would result in one adder and one subtracter; whereas by moving the subtraction one cycle later we could get away with just one dual-purpose unit. Thus in RTL synthesis, the artifact would roughly double the silicon cost. We need to remove such artifacts from the internal description so that scheduling will not be needlessly constrained.

The usual representational mechanism for this is some form of CDFG; the CDFG for the example above is shown in Fig 2.2. The CDFG consists of *nodes*, *edges*, and connections between nodes and edges. The CDFG is very like a netlist; it is not a simple graph because an edge (like a net) can drive many node input 'pins'. Edge $j$ in Fig 2.2 is such an edge. The horizontal lines represent cstep boundaries; i.e. a clock cycle corresponds to the space between adjacent pairs of horizontal lines.

## CDFG nodes

In BC there are five basic kinds of nodes: data nodes, conditional nodes, hierarchical nodes, place holder nodes, and loop control nodes. *Data* nodes represent arithmetic and logical operations, such as addition. They correspond to operators of the source HDL, such as the '+' operator, and under some circumstances to function calls.

Data nodes are further subdivided into *synthetic* nodes, which can share hardware; *random logic* nodes, which cannot; *patch boxes*, which represent bit and field selections, constant sources, and concatenations; *memory reads* and *memory writes*, which represent memory accesses; and *I/O reads* and *I/O writes*, which represent reads of and writes to ports and signals (Verilog has no direct counterpart to the VHDL `signal` construct; a similar result is achieved by using non-blocking writes).

*Conditional* nodes correspond to **if** and **case** and Verilog ?: constructs. Conditional nodes are further divided into *split* and *join* nodes.

A split node represents the 'top' of the conditional, and a join node represents the 'bottom' where the conditional comes back together. Alternatively, a join is analogous to a mux, while a split defines which input of the mux will be selected.

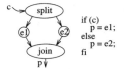

*Hierarchical* nodes contain other nodes and edges. Hierarchical nodes correspond to loops and function and procedure (Verilog **task**) calls.

*Place holder* nodes sometimes represent forward skips in the state graph and sometimes just provide connection points in the CDFG. For the most part these will be invisible unless no schedule can be constructed, at which point they will be listed in reports describing the neighborhood of the CDFG where the contradiction first became apparent. Place holders that represent forward skips are also called *exclusive* place holders: they prevent other operations from occurring during the skipped intervals, and so exclude other operations from the csteps being skipped.

*Loop control* nodes, further subdivided into *loop begin*, *loop end*, *loop exit*, and *loop continue* nodes, denote boundaries of loops.

The loop continue deserves special mention, as its interpretation is not what one might infer from its name. It is the point from which control is passed back to the beginning of the loop. Thus a continue represents the 'bottom' of a loop, and each loop has exactly one continue node. The loop end node, by contrast, represents the latest extent of the loop's scope.

```
        C                      Verilog
                           begin: collar
loopb:  /* loop begin */     forever begin: loop
    if (c1) goto out;           if (c1) disable collar;
    if (c2) goto cont;          if (c2) disable loop;
  cont:  goto loopb;
/* loop end */               end
out:  ...                  end
```

The loop end can actually be later than the loop continue in cases where there is a loop exit pathway containing more states than the 'normal' loop body. The way a VHDL **next** (Verilog **disable**) is represented is with an exclusive place holder (skip) forward to the loop continue. A VHDL **exit** or a C **break** would be represented as a skip forward to the loop end.

## Chaining, multicycling, and pipelined operations

A node will most commonly be scheduled in a single clock cycle, but there is no limit to the number of nodes that can be scheduled in a single cycle, and a node can also be scheduled in multiple clock cycles.

Fig. 2.3 shows two clock cycles, $i$ and $i + 1$. A pair of *chained* additions is shown on the left. The delays of these additions are such that the cascaded pair

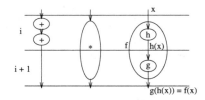

**Figure 2.3.** Chained, multicycle, and pipelined operations

of operations can be performed in a single clock cycle. The command shown below disables all chaining; use it with caution, because even trivially achievable chains will be broken. Use the **set_cycles** (*q.v.*) command to control individual chains.

> bc_shell> bc_enable_chaining = false

A single *multicycle* multiplication is shown in the center of Fig. 2.3. This operation is combinational and has a propagation delay of nearly twice the clock cycle; hence the operation must begin in cycle $i$ and end in cycle $i + 1$. One consequence of this is that the inputs of the multiplier must be registered, as must all control inputs of any muxes that drive these inputs. Otherwise glitches on the control lines may cause bad results to be present on the output when it is finally strobed. But the need to register the inputs and mux control lines means in turn that the control FSM must commit to the multiplication in control step $i - 1$; otherwise there would not be time to load the stabilizing registers.

Thus if the multiplication is in some way conditional (e.g. it is inside a loop with a conditional exit), then an extra control step must lie between the evaluation of the conditional and the beginning of the multiplication. This *latency penalty* is one reason why multicycling is often best avoided.

Here is a command that forces all operations to be scheduled in one cycle, regardless of the results of timing analysis. Use it with caution: it can create situations that logic optimization will have to work very hard to correct.

> bc_shell> bc_enable_multi_cycle = false

On the right of Fig. 2.3, we see a *pipelined* operation $f$. The operation $f$ has been broken down into two operations $g$ and $h$, such that $f(x) = g(h(x))$; $g$ and $h$ each takes less than one clock cycle to complete, and a register is placed between the output of $h$ and the input of $g$. Pipelining operations is accomplished either automatically or by implementation directives. Multiplications and memory operations are particularly common choices for pipelining, but any combinational function and many sequential ones can be pipelined. Note that instances of $f$ can begin in every cycle, with a total time to output of two cycles for each instance. A pipelined component is usually somewhat more expensive than a multicycle component, because of the internal nets cut by retiming; but it is also capable of processing twice as many inputs in the same time, and it does not have the latency penalty.

## CDFG edges

There are two basic kinds of edges in BC: data edges and precedence edges. *Data* edges represent data values, such as $j$ and $c_1$ in Fig. 2.2. The edge $j$ represents a vector of bits; the edge $c_1$ represents a single bit.

*Precedence* edges represent ordering and control, as in the dashed edges $t$ and $f$ of Fig. 2.2. These indicate (1) that the addition and subtraction nodes depend on the result of the conditional, i.e. that one or the other will occur, and (2) that there is an ordering relationship, i.e. that the conditional must be evaluated before one of the arithmetic operations can be decided upon. Precedence edges are also used to represent constraints; for example, the user might constrain one operation to occur at least three cycles after another. This would be represented by a precedence edge with the numeric parameter 'three cycles'.

```
bc_shell> set_min_cycles 3 -from sub1 -to add3
```

In most cases, precedence edges behave as one would expect: they express that one thing should occur before another, or at least concurrently with it. However, there is one case where they can be ignored by the scheduler: where the precedences are from a split to another node. In other words, it is possible, when a conditional operation has no observable side effect, to perform the operation in advance of evaluating the conditional upon which it depends.

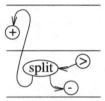

The result of a precomputed operation is saved in a register. When the result of the conditional becomes known, 'untrue' precomputed data, i.e. data that would not have been computed if the precedences had been respected, is discarded; 'true' precomputed data is kept. This is called *speculative execution*. Speculative execution can be used to reduce the latency of a series of conditional operations; it can also increase hardware cost.

Speculative execution is normally disabled in BC, because it radically increases the size of the scheduling search space, and hence scheduling run time; and so the BC team felt it was preferable not to make it the default. Precedence edges that can be ignored in speculative execution are called *weak* precedence edges; all others are *strong*. If speculative execution is turned off, then all precedence edges are strong. Note that it is possible to turn on speculative execution, but still manually constrain individual operations to occur after their conditionals. Here is the command that enables speculative execution: use it with care.

```
bc_shell> bc_enable_speculative_execution = true
```

## Templates

The precedence and data arcs described above all give minimum length constraints. Hence they cannot be used to describe a maximum allowable duration between nodes. Furthermore, there is the problem of pre-scheduled pieces of CDFG. Such a piece is a collection of operations with a rigid timing relationship between its elements; they are allowed to move, but only as a group.

The representational mechanism chosen to represent these and other similar things is the template [11]. A *template* is a rigidly constrained array of slots that can contain place holders and/or other nodes; each slot is exactly one cstep after the last. The template is allowed to move in time as a unit, but the operations in its slots are not allowed to move between slots.

Thus a constraint on the maximum duration between two operations becomes a template, two precedence edges, and two place holders: the first operation must happen after a place holder in the first slot of the template, and the second operation must happen before a place holder in the last slot.

You will not normally see templates unless you have specified an unschedulable design; then the error report may make use of templates to tell you what is wrong. If that happens, just remember that the template is a rigid collection of objects that constrains its members to fall into a rigid timing relationship; and that somehow the rigid timing relationship is participating in a violation of your constraints.

## 2.2.2  Scheduling

BC attempts to minimize hardware costs within the user's timing constraints; so, for example, two additions can be performed on a single adder if they occur in different csteps, or if they are on mutually exclusive branches of a conditional. BC also attempts to minimize the cost of registers. The register cost is bounded from below by the largest sum of bitwidths of the data edges crossing any cstep boundary. The reasoning here is that such data edges represent live data that is generated in one cstep and consumed in another; hence a register of at least the same bit width as the edge is needed.

Scheduling maps CDFG nodes to csteps, as described above. There are many possible ways to do this: see, for example, [6], [9], [12], [15], and [18]. The best general overview I know of is that of [14]. Here is an example of an abstract scheduling algorithm[1].

```
schedule:
   Steps = { 0, 1, ...  Max }
   Unscheduled = { the set of schedulable CDFG nodes }
   while (Unscheduled is not empty) {
      choose Op ∈ Unscheduled such that Op is the most important
      choose t ∈ Steps such that
         (Op, t) is correct and has minimum cost
      schedule Op in t
      remove Op from Unscheduled.
   }
```

---

[1] This is a very simple example. BC uses a much more sophisticated technique; describing it in detail would require an entire book in itself.

The set **Steps** is the set of allowable control steps (csteps), which are integers ranging between 0 and whatever largest allowable time the user has specified. The set **Unscheduled** is initially the set of all schedulable CDFG nodes. In each pass through the loop, an unscheduled node **Op** is first chosen, then scheduled, i.e. assigned to some cstep **t**, and then removed from **Unscheduled**. We choose the most important node remaining in **Unscheduled** in each step; then we find the cheapest step available to **Op**. Common criteria for the choice of **Op** are

**ready** operations, whose operands have already been computed can be scheduled in the current cstep.

**force** is a model of expected numbers of operations over some set of csteps; choose the operation that has the least force exerted upon it.

**highest implementation cost** the operation whose cost is highest should be scheduled first, in the hope that it can use resources that are already committed.

**mobility** operations for which there are few legal choices of **t** should be chosen before operations that have more freedom. Thus operations with critical latency would always be scheduled before noncritical operations.

The analysis of the best node to schedule next can be based on a weighted sum of these and other factors.

The analysis of the set of legal csteps avaliable to a node can become complex. Typical legality considerations for node **Op**, being assigned to a cstep **t**, are

**data inputs** of **Op** must be available or potentially available in **t**.

**consumers** of **Op**'s results cannot be made too late.

**clock cycle** limits cannot be exceeded, e.g. by excessive or inappropriate chaining of **Op** or its neighbors.

**hardware limitations** set by the user cannot be exceeded.

**timing constraints** set by the user cannot be violated.

The analysis of which of a node's legal scheduling options to take can take many forms. One way to do it is to evaluate a weighted sum for each option; then take the option with the lowest cost. Factors in such a weighted sum might be

**earliest possible** is a criterion used in conjunction with the operands-ready operation selection criterion. This gives us a classic and efficient method known as *list scheduling*.

**minimum force** is used in conjunction with the force criterion for operation selection. Place the operation in the cstep where force is minimized; this gives us the very effective method known as *force-directed* scheduling [18].

**resource availability** is a major factor, especially with expensive operations like multiplications. If there are operations of the same type as Op in t, and fewer operations of the same type in cstep u, then u is preferred, because no new hardware will be needed.

**other nodes' freedom** should not be reduced. This class of criterion can become very complicated, using varying degrees of lookahead and backtracking; however, some variations have been shown to give good results.

**register cost** can be reduced by scheduling operations to minimize the number of live data edges crossing the most crowded cstep boundary.

Additional complexities are introduced by the presence of conditionals, loops, jumps, multicycle and pipelined operations, memory operations, loop pipelining, and random logic; by the use of operands of varying bitwidths, by the availability of implementation options; and by the needs of other downstream tools.

Note also that the scheduling problem can be overconstrained. For example, it is the practice in some methodologies to set delay constraints for logic synthesis to zero, forcing synthesis to come up with a high-speed design. This gives the user of such a methodology a rough idea of what can be achieved. The use of an analogous zero-delay constraint in scheduling, however, merely guarantees that no solution will exist. The lesson: either specify achievable constraints, or be prepared to iterate. Iteration at the level of scheduling is very fast compared to logic synthesis, so it is a reasonable exploration strategy.

```
bc_shell> schedule -effort zero
```

## BC schedules bottom-up

BC schedules nested loops and function and procedure calls in a bottom-up manner. Recall that these are represented using hierarchical CDFG nodes. The hierarchy is scheduled leaves-first: first the innermost loop or function call is scheduled, then the next innermost, and so on until the process (outermost) loop is scheduled. As each level is completed, it is inlined, i.e. replaced by its scheduled contents. The contents of a loop are bundled in a rigid timing relationship, using a template, at the time they are inlined. Thus the operations of an innermost loop will be scheduled as a single unit at the next upper and succeeding levels.

Reports of unsatisfiable constraint systems will often include references to templates because of the bottom-up scheduling strategy. This usually happens after an inner loop $L$ has successfully been scheduled and then inlined. Then, as the loop containing $L$ is scheduled, it becomes apparent that the constraints on $L$'s boundaries from outside $L$ are inconsistent with the template that represents the schedule of $L$: and the problem is reported in those terms.

## 2.2.3   Allocation

Following scheduling is the step known as *allocation*. In this step the scheduled operations and edges are mapped to particular hardware resources, which must first be established. Here the number of resources of a given type is not really the question; that has been determined by the schedule. If, for example, there are four multiplications in some cstep, then you need at least four multipliers. What happens in allocation is the mapping of multiplications to specific multipliers; this has implications for both speed and cost. For example, you might already have assigned the operands of a multiplication $M$ to registers R1 and R2, and R1 and R2 might already be supplying operands to a multiplier M1 in another cstep. In that case, assigning $M$ to M1 would not add any new interconnect requirements.

Allocation also has clock speed implications. There are different ways to allocate chains of operations; some have profound effects on how hard logic optimization has to work (and so how many gates it must use) to meet the clock period constraint. For example, consider Fig. 2.4.

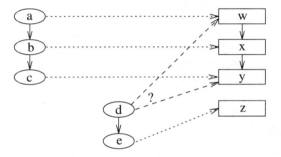

**Figure 2.4.** Effect of allocation on clock period: false path

In Fig. 2.4 we see two chains of operations, $a - b - c$ and $d - e$, on the left, and a set of four hardware boxes on the right. Assume each of the four boxes can perform any of the operations. Suppose also that the two chains are scheduled in different csteps. A chained path from operation $a$ to operation $b$ means that there must be combinational logic (typically multiplexers, etc.) that can steer data from the output of the box $w$ that implements $a$ to an input of the box $x$ that implements $b$. A similar argument holds for the path from the box $x$ to the box $y$ that implements $c$. This interconnect is shown using solid arrows on the right side. Now observe what happens to the path length if we choose to map $d$ to $y$ (lower dashed arrow). The chain $d - e$ means we must construct a logic pathway from $y$ to $z$. The upshot is that we have a combinational pathway that begins with registers feeding $w$ and ends with the registers fed by $z$.

This is a *false* combinational path; the two chains are mutually exclusive. But false paths still cause difficulties for logic synthesis: which will normally respond by trying to make the false path fit into the clock cycle. This means that the logic in the false path will probably end up being needlessly expensive, as logic synthesis tries to achieve a path delay equivalent to 75% of what is really required.

A better allocation maps $a$ and $d$ to the same box (upper dashed arrow of Fig. 2.4). This eliminates the false path, thus saving both delay and area. False cycles are also a problem for logic synthesis; if we were to map $e$ to $w$ we would get such a cycle.

There are many allocation algorithms in the literature; see [14] for a good overview. Allocation algorithms can be divided into three classes: those that perform operator allocation only, those that perform register allocation only, and those that do both. A simple allocation algorithm that does both might look like this. Again, this is not exactly how BC works; BC's actual algorithms are very complex and would needlessly bog us down in details.

```
Unallocated = Operations ∪ Values
Resources = {}
while (Unallocated is not empty) {
    choose U ∈ Unallocated
    construct S = { R | R ∈ Resources and
                        implements(R, U) and
                        free(R, time(U)) }
    if (|S| = 0) add a new R to Resources and to S
    if (|S|= 1) { assign(U, R) }
    else {
        choose the best R from S
        assign(U, R)
        }
    remove U from Unallocated
}
```

This algorithm does allocation in a greedy order, mixing data edges and operations as circumstances permit. The usual strategy for choosing an operation or data edge to be allocated, or a resource to assign, is a weighted sum of factors. Some factors that are taken into account in choosing an operation or edge are

**cost** to implement a suitable resource. Generally speaking, it is better to allocate the more expensive operations and data edges first.

**critical path:** operations and their operands that are likely to affect the clock period should be given the first choice of resources.

**interconnect:** operations and their operands should be clustered so that interconnect can be minimized.

Similar kinds of heuristics can be used to choose which resource to assign to an operation or data edge. For example, if two resources capable of performing an operation `Op` are available during the cstep of `Op`, then we should choose the one that most closely matches `Op` in functionality and bitwidth. As discussed above, it's necessary to avoid false paths and cycles; and the user may have placed allocation constraints on the design.

```
bc_shell> set_common_resource op1 op2 op3 -min_count 2
```

### 2.2.4  Netlisting

The output of BC goes directly to logic synthesis. Hence the scheduled and allocated design must be *netlisted*; this takes the following form.

1. Random logic networks and hardware requested by the user are instantiated.

2. Registers, operators, memories, and so on are instantiated, using the allocation as a guide.

3. Multiplexers, nets, and other connectivity hardware are constructed to connect up the elements of the datapath.

4. The whole is connected to ports and signals.

5. Status and control points are recorded for later hookup to the control FSM's inputs and outputs.

This process is straightforward and does not involve any particular optimization steps; the netlist will be optimized by Design Compiler in any case.

The RTL netlist produced can be viewed using the Design Analyzer. In early versions of BC, this view can be a little misleading if it is done before logic optimization, because the final stages of operation sharing are actually performed as a preliminary step of optimization. Allocation forces a particular sharing, in the same way that a user of RTL synthesis might; but Design Compiler is used to perform the final multiplexing and implementation selection. This is not a fundamental feature; it was simply more convenient to do it this way. The BC team may in the future decide to do the sharing and implementation selection before logic optimization, in which case the pre-optimization view shown by Design Analyzer will show the shared design rather than the unshared design[2].

### 2.2.5  Control FSM

Once the netlist has been constructed, it becomes possible to say what multiplexers and other control points exist, so an FSM can be constructed to drive the control points. This occurs in the following steps:

---

[2] The veteran user of Synopsys's HDL Compiler will recognize the distinction between viewing a design immediately following elaboration and viewing it after compilation; the sharing has not been done, so it looks as if each addition, etc. of the source has its own hardware, whereas in fact these additions, etc. may be shared on a smaller number of hardware resources.

1. A state graph is constructed. This consists of states and arcs that connect the states.

2. A set of control *actions* are constructed. Each of these drives a particular control point; some points (e.g. register load enable pins) will be driven by many actions.

3. The actions are *annotated* onto the transitions of the state graph.

4. Status inputs are mapped from the scheduled CDFG onto particular nets of the datapath. These nets may be either the Q pins of status registers, or the outputs of the hardware that generates the status.

5. The netlist is augmented with a new component, the *control unit*, which is an FSM. The inputs of the control unit are connected to the status signals, and the outputs are connected to the control points.

6. A state table is constructed and attached to the FSM for the consumption of FSM Compiler during compilation. Alternatively, a gate-level FSM can be constructed for you by BC.

### States and csteps

The states of the FSM loosely correspond to the csteps of the schedule. There are two reasons for the lack of a one-to-one correspondence of csteps to states. First, a loop always has one more cstep than it does states; and second, mutually exclusive loops generate disjoint state subgraphs corresponding to the same csteps.

That loops have an additional cstep can be seen by considering the constraints that BC imposes on a loop's CDFG. These constraints state that: first, the loop end must be at least one cstep after the loop begin; and second, the loop exit must be at least one cstep after the computation of the condition that activates the exit. Here's an example.

```
while (cond) begin          while cond loop
    @(posedge clk);             wait until clk'event
                                        and clk = '1';
end                         end loop;
```

This loop has only one state; every time the clock ticks the FSM either goes back to that state or it exits from the loop. But it has two csteps: one for the loop beginning node and one for the loop end node.

There is always at least one clock edge inside a loop; otherwise there would be a combinational feedback cycle and the circuit would not be well-formed. So the two csteps become a single state.

The second reason there is no one-to-one mapping of csteps to states is that loops that are concurrent but mutually exclusive are conceptually difficult to implement on a single set of states due to the disappearing csteps; so loops like the next two are implemented as separate state sequences that have csteps in common.

```
       Verilog                                    VHDL
   if (c1)                                     if c1 then
     while (cond) begin:L1                       L1:  while cond loop
       @(posedge clk);                              wait until clk'event
       @(posedge clk);                                        and clk = '1';
     end                                            wait until clk'event
   else                                                       and clk = '1';
     while (cond) begin:L2                        end loop;
       @(posedge clk);                          else
       @(posedge clk);                            L2:  while cond loop
     end                                            wait until clk'event
                                                             and clk = '1';
                                                  wait until clk'event
                                                             and clk = '1';
                                                end loop;
                                              end if;
```

After control synthesis the netlist is written out in a format that can be viewed, reported upon, or compiled. Depending on the user's goals and current objectives, this may occur many times before logic optimization is finally invoked.

### 2.2.6  Invoking the scheduler

Now that the internals of scheduling have been described, it is time to return to the user's view of scheduling. The BC **schedule** command automatically invokes the tasks of timing (if it has not already been done), scheduling, allocation, netlisting, and control unit synthesis. By setting the **effort** level appropriately, the user can choose between a quick and dirty design (**-effort zero**) for estimation, or a higher-quality design with a longer run time (**-effort high**). The **io_mode** parameter controls the way I/O scheduling constraints are inferred; the three I/O modes of BC are described in chapter 4.

> bc_shell> schedule -effort med -io_mode super

BC's default optimization goal is to minimize the cost of the hardware needed while meeting timing constraints and minimizing unconstrained time intervals. Timing constraints may be explicit, i.e. created by commands such as **set_cycles**, or implicit, i.e. derived from the HDL text.

> bc_shell> set_cycles 5 -from_begin loop1 -to_end loop1

BC can also perform *resource-constrained* scheduling. In this mode unconstrained time intervals are stretched as needed to meet resource constraints. Resource constraints are specified by creating groups of operations, and assigning a resource count to each group. This is the number of resources used for that group, unless timing constraints cannot be met within the resource constraint, in which case more resources will be added. That is, timing constraints always dominate.

> bc_shell> set_common_resource op1 op2 -min_count 1

| cycle | adder1 | addsub | iport | oport |
|-------|--------|--------|-------|-------|
| 0     | add1   | --     | --    | --    |
| 1     | add2   | --     | read  | --    |
| 2     | --     | sub1   | --    | --    |
| 3     | --     | --     | --    | write |

**Figure 2.5.** Reservation table

| state | condition | next | actions |
|-------|-----------|------|---------|
| s0    | ---       | s1   | add1    |
| s1    | 0--       | s1   | read    |
| s1    | 1--       | s2   | add2    |
| s2    | ---       | s3   | sub1    |
| s3    | ---       | s1   | write   |

**Figure 2.6.** State table and graph

## 2.3  BC Outputs

After the schedule command has returned the user can ask for reports. These reports describe the schedule, allocation, and cost parameters of the design. This allows a user to explore the design space without having to do gate-level compilation. For example, the effects of varying the number of ports on a memory or the latency of a pipeline could be explored in order to find the best cost/speed tradeoff.

```
bc_shell> report_schedule -operations -variables
```

The schedule and allocation can be viewed by means of reports that generate reservation tables; a simplified reservation table is shown in Fig. 2.5. Here, the rows represent cycles, and the columns represent resources. For example, the resource **adder1** is an adder. The reservation table in effect tells us what each resource is doing at what time. This is useful for arriving at a general idea of what the resources are, when they are used, and what they do.

Another useful report is the FSM's state graph. This is shown in both tabular and graphical form in Fig. 2.6. This report is useful when you want to find out the latency between one state or operation and another; the condition needed to take a particular branch; or you want to know the structure of a particular loop.

In addition, the netlist can be displayed, and RTL level VHDL or Verilog can be output for simulation purposes. The HDL descriptions output at this stage are technically synthesizable, but they are optimized for simulation; the results of synthesizing these descriptions would not be very good. They are, however, optimized for either debugging or fast synthesis; for example, it is possible to monitor the control unit's state and the operands of synthetic operations when the output is written in debugging mode.

```
bc_shell> write -hierarchy -format vhdl -out mydesign.vhd
```

Simulation at this stage is an important function because it allows the user to check that the behavior of the scheduled design is acceptable. This helps build confidence in the correctness of the scheduling software, but the primary reason for this step is to check the functionality of the interfaces through which the design communicates with its environment. This is necessary because two of BC's three scheduling modes can change the I/O latency, so we must check whether the design can still talk to the surrounding logic.

The design can then be compiled, i.e. logic synthesis can be performed. This process maps the abstract functionality of the design to technology-specific gates, and optimizes the logic so that it meets cycle time constraints at the minimum cost. A number of strategies can be pursued to optimize the cost and speed of the design at this level; think of logic optimization as providing fine-grained optimization of the design that has been coarsely optimized by scheduling.

```
bc_shell> compile -map_effort medium
```

The design can then be *retimed*. This step moves logic through registers; the net result is that the design's clock cycle time decreases. In *behavioral* retiming, the number of registers is also optimized; by contrast, traditional retiming often results in unacceptable increases in the number of registers in the design. The design can then be incrementally recompiled to optimize the area and/or timing given the new register boundaries.

```
bc_shell> optimize_registers
```

Finally, the mapped and optimized logic is written out in an HDL, netlist, or other interchange format for gate-level simulation, place and route, and other downstream tools.

```
bc_shell> write -format edif -hierarchy
```

## Summary

You should now know approximately what the inputs and outputs of BC look like; you should also know how to invoke and control BC; and you should have a working knowledge of the internal representations and general internal flow of BC. This knowledge will help you to understand the reports that you see, and it will also help you to write good HDL input descriptions. If you would rather see actual BC inputs and outputs first, skip forward to chapter 7.

# Chapter 3

# HDL Descriptions

In this chapter we will discuss a basic style for HDL descriptions as they are constructed for behavioral synthesis input. We will begin by discussing the overall structure of the files that describe the design and its simulation environment; we will then give the details of the part of the HDL that BC actually processes. Particular constructs will be described, with emphasis on the ways in which they are interpreted and translated by BC. Finally, we will turn to the test bench, with a general description of test benches and a particular discussion of some good test bench coding practices.

An important feature of an industrially viable tool such as BC is its treatment of interfacing and time in general. The industrial version must provide a rich and convenient set of mechanisms for dealing with pinouts and timing diagrams. Furthermore, these mechanisms must be compatible with simulator semantics; otherwise nearly all of the power of simulation will be lost.

Thus a major theme in the design of BC was support for simulation and comparison of designs both before and after synthesis; we wanted to preserve the possibility of doing a direct cycle-by-cycle comparison between the simulated behavior of the presynthesis and the post-synthesis designs. This was one of the primary design goals of BC; some of the design decisions whose effects are described in this chapter can best be understood in this light.

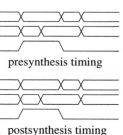

presynthesis timing

postsynthesis timing

HDLs such as Verilog and VHDL support the description of timing in numerous powerful ways. Because we wanted to support simulation both before and after synthesis, it was necessary to design BC so that it would interpret the simulation semantics of an HDL text and synthesize a circuit that would mirror the HDL text's timing in structured and predictable ways.

But this also imposes restrictions on what can be said in the HDL text. Timing diagrams that can't be synthesized, or constructs that could lead to unpredictable or unreliable circuits, must be avoided *in the HDL text*. This chapter describes the results of a dialectic between the HDL's descriptive power and BC's ability to construct the circuits the HDL can describe.

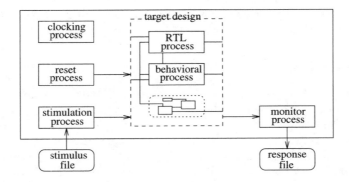

**Figure 3.1.** HDL before synthesis

The usual structure of the HDL files as constructed by a BC user is shown in Fig. 3.1, which shows a hierarchy of simulation entities. The largest box is the test bench, which is the root of the simulation hierarchy. The test bench is a substrate without which the design cannot be simulated; it supplies stimuli and monitors the responses of the design. The test bench shown here is generic; many variations are possible. This one is organized for pre-synthesis use; it has five subdivisions.

The *clocking process* generates a clock signal that drives and synchronizes the other processes and the design.

The *reset process* generates a reset pulse at the beginning of simulation, and occasionally at other times as well. This serves both to initialize the design (which otherwise might get stuck in an unknown state) and to test its response to a reset.

The *stimulus process* generates and delivers inputs to the design. It may be self-contained, e.g. a counter, or it may take input from a file. It often takes synchronizing inputs from the design as well, and may also have monitoring functions.

The *monitoring process* looks at the outputs of the design and logs them to a file; alternatively, it compares the outputs with some 'golden' file or test process, and reports a match or mismatch. For example, if the design computes the square root of its input, then the monitor might compute the square root independently and compare the two results, reporting only that the two numbers agree.

The *target design* is the user's HDL input to synthesis. It can consist of RTL processes (upper), behavioral processes (center), and random logic (lower, dotted box), as well as the internal nets necessary to connect these things together and to the design's ports.

The design should be thoroughly tested at this level before synthesis begins. The main goal here is to be sure that the design works as it is specified, i.e. that its various components are correctly defined and that they all talk to one another and to the test bench correctly. Pre-synthesis simulation is typically much faster than post-synthesis simulation; so it is best to catch as many bugs as possible at this stage. It is also likely that much or all of the test bench will be reused after synthesis; so time spent developing a good test bench and good stimulus/response files is time well spent.

## 3.1  The Design

The design to be synthesized is described as either a VHDL entity and an associated architecture, or as a Verilog module.

```
-- VHDL                                // Verilog
library IEEE;                          module small (clk, reset, ip, op);
use ieee.std_logic_1164.all;            input clk, reset;
use ieee.std_logic_arith.all;           input [7:0] ip;
entity small is                         output [7:0] op;
  port (clk, reset:  in std_logic;      reg [7:0] op;
     ip:  in signed (7 downto 0);       ...
     op:  out signed (7 downto 0));    endmodule
  end small;
```

These declarations create a pinout for the design. The design going into BC must be a single such module or entity; separate designs can be combined after synthesis. With the exception of tristate logic, all of the types and packages of the RTL synthesizable subsets (see Appendix B) of VHDL and Verilog are supported.

In VHDL the ports and signals of an entity both have signal-like behavior. In Verilog, any port or net driven by behavioral or RTL processes (**always** blocks) should be registered; hence the **reg** declaration of the **op** port.

In Verilog, we would simply continue at the ellipsis. In VHDL, by contrast, we must now declare an architecture for the entity given above. The architecture defines the 'guts' of the design for which the entity defines the pinout.

```
architecture be of iir is
  -- type, signal, component, and other declarations
begin
  -- components, processes
  ...
end be;
```

Quite often, a design will just consist of a number of processes, connected together by nets; but there may be additional random logic or other instantiated components as well. These nets and instantiated components are declared at the level of the module or architecture.

## 3.2  Behavioral Processes

BC schedules only processes. Anything outside a process will be left alone by scheduling. Thus you can wrap a process in a jacket of random logic, and the jacket will be preserved through synthesis.

Any number of processes may be present.  A process may be either RTL or behavioral: the difference being that RTL processes are ignored by scheduling.

If there are multiple behavioral processes, they will be scheduled independently of one another; that is, the schedule of one process has no effect on that of another. BC makes no attempt to maintain synchronicity between scheduled processes, or between a scheduled process and an RTL process. This is an important consideration when, for example, the flow of control in a process is such that alternative pathways in the state graph have different numbers of cycles. In that situation the user must provide strobes, ready signals, and so on in the HDL description in order to maintain proper synchronization between the processes.

Synchronization is also an issue when using any I/O mode other than cycle-fixed (see Chapter 4), because in the other modes the I/O timing of one process may be transformed while that of another may not. This means that inter-process communication protocols must be such that they still work if cycles are inserted or removed. Alternatively, the user must add manual constraints to prevent timing changes if timing changes would cause trouble.

Consider a single process to be synthesized by BC. In VHDL this is a **process** without a sensitivity list; we will want to use **wait** statements inside the process to synchronize it with the clock.

```
main:  process
   -- local variables
   variable x:  signed(7 downto 0);
   ...
   begin
     -- behavioral statements
     ...
   end process main;
```

The local variables we declare inside the process will be visible only within the process; conceptually they are mapped to logical registers, which are mapped to physical registers or nets in the allocation phase.

In Verilog, our process is an **always** block. It may have local variables; many users prefer to use global (module-level) variables instead. If a global variable is accessed by two or more processes, it is treated as an I/O port; otherwise it is the same as a local variable. Any local variables of a process, or variables written by a process, must be declared as **reg** variables. Output ports, as declared above, are also declared to be **reg**s if they are assigned to inside the behavioral process.

```
always begin:  main
   -- local variables
   reg [7:0] x; // must be a reg
   ...
   end
```

The process will be compiled to form a loop. That is, if the flow of control ever gets to the end of the process block, it will 'jump' back to the beginning.

## 3.3  Clock and Reset

To describe a clocked sequential system in a realistic way, we need to describe the relationship of the system to the clock. In addition, we need to describe the way reset works, so the system can be initialized. BC supports a single-phase edge-triggered clock, and either synchronous or asynchronous resets.

BC uses clock edge statements to tie its behavioral descriptions to time. Each clock edge statement forces the behavioral process to await the next active clock edge before proceeding with 'execution'. For example, an output write operation can be forced to fall one cycle after another operation (e.g. a port read) by putting a clock edge statement in between the two corresponding HDL statements. BC permits a user to insert any number of clock edge statements in a process. In VHDL, we use the `wait` construct; in Verilog, we use the `@` construct.

```
wait until clk'event and clk = '1';          @(posedge clk);
```

Either rising or falling edges can be used; but edge polarities cannot be mixed in a single process, and only one clock signal is permitted in a process. Different processes can use different edges, clock nets, and frequencies. Process sensitivity lists are not allowed in a behavioral process, and inside a behavioral process only one signal and polarity can be the argument of a `wait` or `@`.

### Resets

To simulate the behavior of a synchronous reset, i.e. one that takes effect on a clock edge, we insert a loop exit (Verilog `disable`) statement following each clock edge. The loop to be exited encloses the entire behavior of the process. This loop (we will call it the *reset loop*) should begin with reset-specific behaviors (we will call these the *reset tail*). After the reset tail the process will normally contain another loop that contains normal mode behaviors (we will call this the *main loop*).

For example, a simple microprocessor would, upon reset, load zeros into its program counter: this would be the reset tail. In the main loop it would perform a fetch-execute cycle.

In the simple CPU shown on the facing page there are only two clock edge statements; one is in the reset tail, and one is between fetch and instruction decode. Each and every clock edge in the process is immediately followed by a conditional loop exit. When reset goes true, the state immediately following the next clock edge will be the first state, in which the PC and SP are initialized.

In order to capture the usual idea of how a machine should behave when a reset occurs, there must be a conditional `if (reset ...` branch to the process end after every clock edge statement. That will simulate reset in a reasonable way: and BC will infer a synchronous reset.

```
      -- VHDL                            // Verilog
main:  process                     always begin:  reset_loop
begin                                 // reset tail
   reset_loop:  loop                  pc = 16'h0000;
      -- reset tail                   sp = 16'hffff;
      pc := (others => '0');          @(posedge clk);
      sp := (others => '1');          if (reset == 1'b1)
      wait until clk'event               disable reset_loop;
           and clk = '1';            forever begin:  main_loop
      if (reset = '1') then             // normal mode behavior
         exit reset_loop;               instruction = memory[pc];
      end if;                           @(posedge clk);
      main_loop:  loop                  if (reset == 1'b1)
         -- normal mode behavior           disable reset_loop;
         instruction := memory(pc);     case (instruction)
         wait until clk'event              8'h3E: ...
              and clk = '1';               8'h9C: ...
         if (reset = '1') then          endcase
            exit reset_loop;         end
         end if;                   end
         case (instruction) is
           when "00111110" => ...
           when "10011100" => ...
         endcase
      end loop main_loop;
   end loop reset_loop;
end process main;
```

If one reset branch is missing, or if all of the reset branches are not identical, they will not be treated as reset branches and scheduling will be overconstrained. When BC detects a properly formed set of reset statements, it reports:

> `A global synchronous reset has been inferred.`   (HLS-39)

If it does *not* say this, you should check the source HDL, because the quality of results obtained could be quite poor if, e.g., one reset branch was omitted. You will also see an error message if any of the loop exits that are 'left over' are nonlocal.

Some users don't like to put in the loop exit after every clock edge statement. It does, after all, make the code more verbose. Unfortunately it is the only way to make the design respond to a reset in simulation. Verilog users tend to use a macro to define a combined clock edge and reset.

```
`define clk_edge                                \
     begin                                      \
        @(posedge clk);                         \
        if (reset == 1'b1) disable reset_loop; \
     end
```

If the response of your behavioral HDL to a reset is not a concern, which is often the case in the early phases of design, then there is no need to include conditional exits after clock edges. However, if you want to simulate the output of BC you will still need a reset; otherwise all of the state and status registers of the synthesized design will be uninitialized and the design may not simulate at all.

The method of getting reset behavior into the scheduled design without including it in the HDL is to use the bc_shell command **set_behavioral_reset**: this results in a reset being created and connected. In order to use this command you must have a reset net or port in the HDL, even though the process never explicitly reads it; otherwise BC will exit with an error message. In this example the name of the reset port is **reset** and the reset is true high.

```
bc_shell> set_behavioral_reset reset -active high
```

Nets that drive no input pins and are not read by any process are automatically deleted during elaboration. Hence the use of an internal net to drive a BC process's reset requires the construction of dummy logic or a dummy port. This dummy logic or port is driven by the reset net, and so protects the reset net from being deleted.

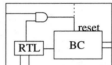

### Asynchronous resets

An asynchronous reset is one that takes effect immediately when the reset pin goes true, regardless of the state of the clock. An asynchronous reset is synthesized in response to the bc_shell command **set_behavioral_async_reset**, which can either be given to bc_shell directly or embedded in the HDL text. This command causes registers built by BC to have asynchronous resets.

If you really need to have the pre-synthesis design simulate an asynchronous reset, the HDL code can be written like this:

```
-- VHDL                              // Verilog
wait until (clk'event and clk = '1')  @(posedge clk
    -- synopsys synthesis_off        // synopsys synthesis_off
    or (reset'event and reset = '1')   or posedge reset
    -- synopsys synthesis_on         // synopsys synthesis_on
    ;                                  );
if (reset = '1') then                if (reset == 1'b1)
    exit reset_loop;                     disable reset_loop;
end if;
```

Most users don't need the level of modeling precision that this construct implies; and it decreases readability. In Verilog, I normally use a macro for this kind of thing; in VHDL, one can use **m4** macros.

## 3.4   I/O Operations

Recall from Section 2.2.1 that there are special CDFG nodes for I/O reads and writes. These are inferred from the source HDL whenever the behavioral process references a variable (in VHDL, a `signal`) external to the process. In VHDL, this can be a port of the entity or a signal of the architecture; in Verilog, it is a port of the module or a variable declared at the top level of the module and referenced by some other process. When such a variable or signal is written inside the behavioral process, an I/O write is inferred; when it is read, an I/O read is inferred. Fig. 3.2 shows examples of reads and writes.

Observe that there are two reads created in the first cycle. These are in a sense 'duplicates'; they read the same port during the same cycle. BC will not consolidate them in a single read. The reason for this is that in two of the BC I/O scheduling modes it is possible for the time between explicit clock edges to be varied during scheduling, rather as if new clock edges had been inserted into the HDL text. Keeping the reads distinct allows them to be scheduled at different times within the stretched time interval. This in turn can be used to reduce the overall resource requirement, because the earlier of the two reads can feed an early operation, and the later read can feed a later operation: so it is unnecessary to save the data value obtained in the early read while waiting for the later operation's assigned cstep. If you want to have just one read, that is easy enough: just re-use the variable `v1` to feed both computations. But a read operation doesn't cost much in terms of hardware, and there is a real gate count associated with saving a data value.

The write to `sig` looks just like a write to an output port. The '`<=`' construct behaves the same way in both languages, for both internal signals and ports: the signal will transition at the next clock edge, and then hold its value until the clock edge following the next assignment. This has a corollary: BC registers all output ports and signals written by '`<=`'.

Using '`<=`' when you don't need the registered signal semantics will create extra storage and possibly latency, because all such writes are treated as I/O writes, and come under several rules that affect I/O writes. Thus if you don't need to communicate outside the process, you should normally avoid the use of '`<=`'.

There is, however, one situation where a signal write to a local variable is useful. The signal write creates the equivalent of an I/O write: and any signal is also read as if it were an I/O port. What's more, the reads and writes of that signal come under all of the rules concerning I/O scheduling.

This means you can force a particular variable timing and register usage, simply by using a signal write. The most interesting case of this that I have seen was a large, complex computation that needed to be done in several clearly-defined stages: there was no I/O inside the computation. This resulted in a long run time, because the scheduler had to explore such a large solution space. So the designer defined some variables that were produced in each stage of the computation; and treated these variables as I/O ports, thus locking down both the timing of the stages and the usage of the variables. This reduced the search space, constrained the schedule,

```
-- VHDL
entity small is
   port(din  : in signed (7 downto 0);
        dout : out signed (7 downto 0);
        clk  : in std_logic);
end small;
architecture b of small is
   signal sig : signed (7 downto 0);
begin
   behavioral: process
       variable v1, v2 : signed (7 downto 0);
   begin                 -- no loops or resets! A strange design.
       wait until clk'event and clk = '1';
       v1 := din; -- this creates a read
       v2 := din; -- this creates another read!
       wait until clk'event and clk = '1';
       sig <= v1; -- creates a write
       wait until clk'event and clk = '1';
       dout <= v2 + sig; -- creates a read and a write
       wait until clk'event and clk = '1';
   end process behavioral;
end b;


// Verilog
module Small (din, dout, clk);
    input clk;
    input [7:0] din;
    output [7:0] dout;
    reg [7:0] dout, sig;
    always begin: behavioral    // No loops or resets! Odd.
       reg [7:0] v1, v2;
       @(posedge clk);
       v1 = din;    // creates a read
       v2 = din;    // creates another read!
       @(posedge clk);
       sig <= v1; // creates a write
       @(posedge clk);
       dout <= v2 + sig; // create a write and a read
       @(posedge clk);
    end
endmodule
```

**Figure 3.2.** Variable and I/O reads and writes

and incidentally made testing, etc. much easier, by giving exact control over the contents and timing of the 'registers' that BC built to stabilize the 'output signals'.

In Verilog the situation is a little more general, because any **reg** variable of the module's declarations can be written either using a blocking ('=') or a nonblocking ('<=') write. BC allows you to use either, as long as the variable is not a port, and as long as it is neither read nor written by another process. If it is, you must use the nonblocking write. You should usually use a blocking write for a local variable; again, the nonblocking write creates a dedicated register and a set of timing constraints that you may neither need nor want.

The write to **dout** is as you would expect: it is a port, so the '<=' construct is used in both languages. This implies a registered output that holds its value until the next time you write it. It generates a CDFG write operation, that in this case is fed by a CDFG addition, which is in turn fed by the value currently in **var** and a CDFG read operation from the signal **sig**. Thus a read can drive a write in the same cycle. Notice, however, that the result will not appear until the next clock edge, because all outputs are registered.

BC assumes that data on input ports transitions on clock edges, i.e. that all drivers of BC input ports are registered. This mimimizes troubles with setup and hold times.

It is as well to remember, in this context, that neither **v1** nor the addition operations are necessarily preserved in the form they have here. Scheduling may move the operations in time, and allocation may share registers with other values. However, the I/O operations are much less volatile: they will only be scheduled in very structured ways (e.g. in the cycle-fixed I/O mode of Chapter 4, their timing is never changed at all), and their locations (i.e. the ports and signals) are never changed under any circumstances. These restrictions guarantee that your ports won't be changed around, and that you can control the ways in which their timing will be shifted from one cycle to the next. The exact ways in which you can control I/O timing are described in Chapter 4.

## 3.5  Flow of Control

Behavioral Compiler supports most of the ordinary flow-of-control constructs of VHDL and Verilog. This includes **for** and **while** loops, nonterminating loops, if-then-else statements, the VHDL **elsif** construct, case statements, functions, (VHDL) procedures, and (Verilog) **task**s.

**Next**, **exit**, and **disable** constructs are supported, as long as they are either exits or continues of the loop in which they are immediately enclosed, or they are associated with a reset. That is, only the loop exits and continues associated with resets may be nonlocal, and other exits and continues cannot be used inside a loop nested within the loop that is exited or continued.

In Verilog the situation is a little more complex because of the semantics of **disable**. The **disable** causes a named begin-end block to be exited. If the block is a loop body, then exiting the block will just continue the loop. Thus in order to exit a loop, you must disable a begin-end block that immediately contains the loop you wish to exit.

```
begin:collar
  forever begin:loop
    . . .
    if (c1) disable collar;
    if (c2) disable loop;
  end // loop
end // collar
```

Loops that have fixed iteration bounds (**for** loops whose iteration boundaries are compile-time constants) are *unrolled* at elaboration time by default. This eliminates the hardware that evaluates the conditional; it can result in either increased or decreased numbers of states; it allows a user to write loops that contain no clock statements; and it allows operations taken from successive iterations of the loop and operations outside the loop to be scheduled simultaneously.

The default must be used with caution when large iteration counts or deeply nested loops are used. Consider the following fragment of VHDL.

```
for i in 0 to 9 loop
    for j in 0 to 9 loop
        for k in 0 to 9 loop
            wait until clk'event and clk = '1';
            -- operations
        end loop;
    end loop;
end loop;
```

Unrolling this fragment would result in a thousand or more csteps and states in the synthesized FSM, slowing down scheduling, which explores the space of scheduling solutions. BC supports the attribute **dont_unroll**, which causes BC to build the loop with an explicit test. For the nested loops shown above, this would result in a state machine that had only one or a few states to implement the loop, instead of a thousand.

The **dont_unroll** attribute is used as follows.

```
  -- VHDL
attribute dont_unroll: boolean;
attribute dont_unroll of Loop_name: label is true;
        .   .   .   .
Loop_name: for i in 0 to 4 loop . . .

  // Verilog
/* synopsys resource foo: dont_unroll = "Loop_name"; */
        .   .   .   .
for (i = 0; i < 5; i = i + 1) begin: Loop_name . . .
```

Infinite loops and loops whose indices must be computed at 'run time' cannot be unrolled at all. Thus, for example, `while` loops, `forever` loops, and VHDL loops with no loop controlling statement cannot be unrolled. Moreover, if a loop contains a conditional exit other than the loop test, it cannot be unrolled. You cannot partially unroll a loop: but you can always achieve the same effect by nesting unrolled and rolled loops.

### Pipelined loops

A *pipelined* loop is one in which successive iterations are performed in parallel. This can substantially improve the performance of your design, and sometimes it can save you gates as well. Don't confuse this with pipelined components: the two do go well together, but they are not the same and do not have to be used together.

Suppose, for example, there is a loop containing five arithmetic operations. In the first cstep of the loop a sample is read; in the last, a result is written. Suppose that the schedule is constructed in such a way that the operations occur at exactly the same times in this piece of HDL and in the scheduled design, perhaps because timing considerations prevent BC from chaining the operations.

```
loop
    a := inputport;
    wait until clk'event and clk = '1';
    b := op1(a);
    wait until clk'event and clk = '1';
    c := op2(b);
    wait until clk'event and clk = '1';
    d := op3(c);
    wait until clk'event and clk = '1';
    e := op4(d);
    wait until clk'event and clk = '1';
    outputport <= op5(e);
    wait until clk'event and clk = '1';
end loop;
```

Suppose also that the operations *op1, ...op5* are sufficiently different that they cannot share the same hardware resources. Then it is necessary to allocate a separate functional unit for each operation, and each functional unit will be used only 1/6th of the time.

We can represent the scheduled loop as a rectangle; it is composed of a stack of six smaller rectangles, each representing a cstep. In the first cstep, the input is read; in the last, the output is written. In between the operations of the loop are distributed, one in each cycle.

The loop has a *loop latency* of six cycles; that is, from the beginning to the end of each pass of the loop is six cycles. The loop has an *initiation interval* of six cycles

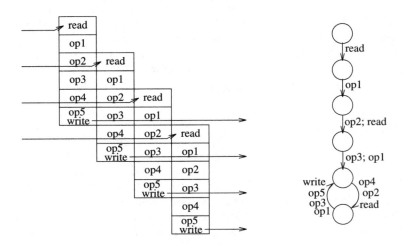

**Figure 3.3.** Pipelined loop: Time/space chart and state diagram.

as well; that is, six cycles fall between successive loop beginnings. Throughput is one sample every six cycles, and we need five operation modules.

Now notice that there is nothing stopping us from overlapping successive passes through the loop. This is the key idea of loop pipelining: that the loop can be restarted before it is finished.

On the left side of Fig. 3.3 we show the same loop pipelined with an initiation interval of two. The latency is still six. This means that once the pipeline fills, there are three iterations of the loop executing in parallel; and that every other cycle a new iteration begins with a new sample being read. Throughput is tripled, even though it still takes six cycles to process a single sample. Moreover, there is no additional operation cost. The component that performs *op1* still only performs *op1*; it is just doing so three times as often. There may be additional storage and interconnect costs, but sometimes loop pipelining decreases cost across the board.

In loop pipelining, the initiation interval can be set to any value that evenly divides the loop latency. For example, an initiation interval of two and a latency of six is acceptable, but an initiation interval of three and latency of seven is not.

The state diagram for the pipelined loop is shown on the right of Fig. 3.3. The state on the top is the entry to the loop; notice the first read on the transition from the first to the second states of the loop. The second pass begins at the third state; notice the data read. At the bottom the pipe is full; in each of the two lowest states, half of the operations of the loop are performed.

There are two ways to create a pipelined loop in BC. The first can be used only in fixed I/O mode; in this mode, the simulated design and the synthesized design have the same cycle-level timing (see Chapter 4). For this example we will return to the little filter $x[k] = ax[k-1] + bu$ of previous chapters. We assume that the clock has a ten-nanosecond period.

```
        -- VHDL                          // Verilog
ppl: loop                            forever begin: ppl
   u := inp; -- read u                  u = inp; // read u
   x := x * a + u * b;                  x = x * a + u * b;
   outp <= transport x                  outp <= #20 x;
          after 20 ns;                     // 2 clock cycles
          -- 2 clock cycles
   wait until clk'event                 @(posedge clk);
          and clk = '1';
end loop ppl;                        end
```

Notice that a transport delay modifier has been used on the I/O write to **outport**. The delay forces the effect of the write to take place two cycles after it otherwise would, but the loop itself has only a single state in fixed I/O mode.

In other words, *the write takes place during a subsequent iteration of the loop.* The amount of delay you insert here determines the latency of the pipeline; the number of clock edges in the loop determines the initiation interval. You have to be careful about this: the latency must be evenly divisible by the initiation interval.

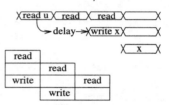

The delay is an integer number of cycles when it is seen by BC; but you have to express the delay in nanoseconds (or whatever other time unit you use) when you write the code. A small error will make no difference; the delay gets rounded to the nearest clock cycle, so you can afford to be off by up to half a clock cycle.

In the other I/O modes (see chapter 4) there is a less rigid relationship between the timing of the simulated and synthesized designs; so an implicit declaration of a pipeline as shown above is neither necessary nor feasible. Instead, the bc_shell command **pipeline_loop** is used.

bc_shell> pipeline_loop ppl -initiation 1 -latency 3

This command can be given in a bc_shell script, or it can be embedded in the source text using the **synopsys dc_shell_begin** and **_end** pragmas. Note that this command's parameters can be modified in the script to achieve a completely different pipeline latency and initiation interval. By contrast, in fixed I/O mode, the same changes would require changes to the source HDL. This increased flexibility is a major advantage of the superstate and free-floating modes over fixed mode.

## Exits from pipelined loops

In early versions of BC the only way to exit from a pipelined loop was to reset the process. We later found a way to generalize this; BC can now handle exits from pipelined loops. This is done by providing states and operations that flush the pipeline after the exit. BC provides different flushing sequences for each unique number of previous loop iterations that can be 'active' at the exit time.

Recall also that BC provides an exact timing match in the fixed I/O mode. This means it is necessary to provide time in the source HDL during which the pipe can be flushed. The time you must provide takes the form of clock edge statements following the loop end. The reason you need these is that the next operation whose schedule can be observed (e.g. an I/O write) cannot be permuted with the operations that occur during the time the pipeline is emptying out after the loop is 'exited'. So there need to be some states after the loop in which no such operations need to happen; the 'extra' clock edges allow BC to provide these states.

Consider, for example, a slight modification of the little filter example. This one uses a **while** loop and the next operation after the loop is an I/O write.

```
    -- VHDL                              // Verilog
ppl: while (cond) loop              while (cond) begin: ppl
    u := inp; -- read u                 u = inp; // read u
    x := x * a + u * b;                 x = x * a + u * b;
    outp <= transport x                 outp <= #20 x;
            after 20 ns;
                -- 2 clock cycles            // 2 clock cycles
    wait until clk'event                 @(posedge clk);
            and clk = '1';
end loop ppl;                       end // ppl
outp <= out_of_order;     -- illegal --  outp <= out_of_order;
```

Something like this in the HDL will give us a pipelined loop with an exit; but unfortunately it is illegal. Consider the events that occur immediately after **cond** goes false. In previous passes through the loop, we committed ourselves to writes that have not yet occurred; these writes are still in the pipe, and will conflict with the write that follows the loop. This results in the I/O operation ordering being permuted. In order to avoid such permuted I/O operations, BC requires that the user add extra clock edges (in fixed I/O mode) or that BC be allowed to add extra clock edges (in superstate and free-floating modes). The trailing states provide time to flush the pipe and thus avoid permuting the writes.

```
ppl: while (cond) loop                  while (cond) begin: ppl
    u := inp; -- read u                     u = inp; // read u
    x := x * a + u * b;                     x = x * a + u * b;
    outp <= transport x after 20 ns;        outp <= #20 x;
            -- 2 clock cycles                   // 2 clock cycles
    wait until clk'event and clk = '1';     @(posedge clk);
            and clk = '1';
end loop ppl;                           end // ppl
wait until clk'event and clk = '1';     @(posedge clk);
wait until clk'event and clk = '1';     @(posedge clk);
wait until clk'event and clk = '1';     @(posedge clk);
outp <= now_in_order;                   outp <= now_in_order;
```

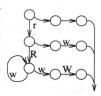

Here is the state diagram for the simple filter with a pipelined `while` loop. Note that lowercase reads and writes go together, as do the uppercase ones. Notice also that the effect of the trailing clock edge statements in the HDL is that the trailing states are created regardless of how many times the loop has been executed; that is, they will be present even if the pipeline has not had a chance to fill.

Note also that if you use a scheduling mode that can automatically add the 'extra' states without requiring them in the source HDL, you will still have to plan for the existence of the trailing states in the synthesized design. That is, you can't constrain the output write to be too close to the loop end, regardless of the I/O scheduling mode you use.

Notice that in the uppermost exit sequence (extending to the right from the upper left state) nothing is done; the exit was taken in the first pass through the loop. Thus no write will occur and no operation inside the loop needs to be flushed. The second exit sequence begins in the second pass through the loop; one previous pass needs to be completed and flushed out of the pipe. This is shown by the presence of `w` on that sequence. Exits occurring after the second pass will use the third exit sequence; then two passes need to be flushed (`w` and `W`). For exits occurring after two passes, the number of queued writes is still always two, because only two previous loop passes will still be active when the exit branch is taken. A loop with longer latency and/or a shorter initiation interval would have corresponding numbers and lengths of loop exit sequences. The best way to compute these parameters without running BC is to draw the state diagram: any other approach risks confusion.

Because of an explosion of possible states, BC disallows exits occurring later than one initiation interval from the loop beginning. For example, if the initiation interval of a loop was five, then no exit could be scheduled later than six cycles from the loop's beginning.

Another thing BC disallows is rolled loops nested inside pipelined loops. An unrolled loop causes no trouble; it is unrolled at elaboration time, and so just becomes part of the scheduling problem. But a rolled loop in principle has a data-dependent number of iterations, which would play havoc with BC's ability to statically determine which csteps of iteration $i$ would be concurrent with which csteps of iteration $j$. This in turn would eliminate most of the benefits of loop pipelining.

## 3.6  Memory Inference

Behavioral Compiler allows the user to specify memories using arrays. These memories must be thought of as consisting of words; each word is accessed as a unit. This allows a convenient and intuitive specification of memories; BC takes care of the details of scheduling memory accesses and controlling the memory's ports, strobes, etc..

```
x = 0;
for (i = 0; ...) begin
  x = x + u[i] * C[i];
  u[i] = u[i - 1];
end
```

```
-- VHDL
architecture beh of mem_dsg is
   subtype resource is integer;
   attribute variables : string;
   attribute map_to_module : string;
   type mem_type is array (0 to 15)  -- number of words
           of signed (7 downto 0); -- word width
 begin
   behavioral: process
      constant Mem1 : resource := 0;
      attribute variables of Mem1:   -- physical memory
         constant is "M";
      attribute map_to_module of Mem1:
         constant is "DW03_ram1_s_d"; -- type of memory
      variable M: mem_type;          -- logical memory
   begin
      .  .  .  .
      M(12) := "00001010"; -- memory write
      outport <= M(2 * x); -- memory read


 // Verilog
 module mem_dsg ( .  .  . );
    reg [7:0] M [0:15]; // declare logical memory
    /* synopsys resource Mem1: variables = "M"
             map_to_module = "DW03_ram1_s_d"; */
    .  .  .  .
    M[12] = 8'h0a;       // memory write
    outport <= M[2 * x]; // memory read
```

**Figure 3.4.** Memory inference

A memory is declared using an array variable and a pragma that maps the variable to an instance of a particular memory in the synthetic library. Fig. 3.4 shows an example of memory inference.

We begin by declaring a resource **Mem1**; this resource has attributes that tie it to a set of variables (here just **M**) and a DesignWare memory (here **DW03_ram1_s_d**) that will be used to implement the memory. The resource **Mem1** represents the physical memory, which is mapped to logical RAM; this is useful in cases where there are multiple array variables assigned to the same physical memory.

If you want an array of dimension two, where each element is a word, then you have to manage one of the array dimensions yourself, without assistance from BC.

The two most important things to remember about RAM are: first, RAM accesses are synthetic CDFG operation nodes; and second, BC makes conservative assumptions about address conflicts. The fact that RAM accesses are synthetic operation nodes means that they can be scheduled and allocated, and that memory parts must be present in your DesignWare library. A memory can be sequential or single-cycle; it can also support pipelined accesses, depending on how the ASIC vendor chose to build it. A memory can also have any number of ports, again depending on the vendor.

Because memories can be scheduled, BC also has to ensure that reads and writes to a single address don't get out of order. An *address conflict* occurs when two accesses to the RAM might touch the same address, and at least one of these accesses is a write. This possibility forces BC to keep the original source code order between reads and writes and between writes and writes. For example, in these fragments the read operation must never be scheduled before the write, because it would see the wrong data if it were.

```
     VHDL                 Verilog
M(14) := 5;          M[14] = 5;
x := M(14);          x = M[14];
```

If both operations were reads, of course, there would be no problem; but no read of a particular cell can move past a write of that cell. If we had the fragments

```
     VHDL                 Verilog
M(14) := 5;          M[14] = 5;
x := M(13);          x = M[13];
```

there would be no reason not to interchange the order of the read and the write; there is no conflict because different cells are accessed. BC won't interchange them, however, unless you take special measures. The reason BC won't interchange these operations is that it does not attempt to prove that the addresses being accessed are in fact disjoint under all circumstances. Disjointness is easy to see in the case where the addresses are constants; but it is much less straightforward where arithmetic expressions and variables must be evaluated.

By default BC creates CDFG precedence edges from all reads of an array to neighboring writes to the same array, and between all neighboring pairs of writes. If you believe that there are no memory conflicts, or if for some reason your algorithm can tolerate permuted accesses, there is a way around this:

```
bc_shell> ignore_memory_precedences -from op1 -to op2
```

This command says that the user believes that the two named memory operations do not conflict, and that BC should not construct a memory precedence constraint between these two operations. Be careful when you use this command: you have made yourself responsible for ensuring that no out-of-order accesses occur.

Another case where memory constraints can be overly conservative is in pipelined loops. In a normal loop, the memory reads and writes of each pass through the loop

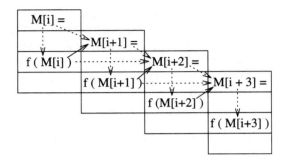

**Figure 3.5.** Pipelined loop with memory accesses

are 'insulated' from the reads and writes of other passes by the temporal ordering imposed by the loop. That is, the iterations of the loop are strictly ordered just by the way the loop works. But because a pipelined loop has simultaneously executing iterations, there are cross-iteration memory constraints that must be taken into account so that the ordering property of the original HDL is preserved in the synthesized design. This can be seen in the following fragments and the accompanying pipeline schedule (Fig. 3.5).

```
for (i in 0 to Msize) loop
   M(i) := inport;
   outport <= transport f(M(i)) after 20 ns;
   wait until clk'event and clk = '1';
end loop;

for (i = 0; i < Msize; i = i + 1) begin
   M[i] = inport;
   output = #20 f(M(i));
   @(posedge clk);
end
```

Notice that the memory accesses in different iterations touch different cells. But because BC does not attempt any inference of this kind, this is not apparent to BC's algorithms; so BC installs precedence arcs in the CDFG that constrain the accesses in consecutive passes through the loop. These are shown as dotted and solid arrows in Fig. 3.5.

The precedences make it impossible to pipeline the loop with the given initiation interval. Observe the solid arrows going upward and to the right. In the unpipelined loop, the memory operations of one pass through the loop all precede the memory operations of the next. In the absence of the inference that M[i] and M[i + 1] are in fact disjoint, the memory operations of the second loop must be scheduled chronologically after those of the first loop; but the solid arrows of Fig. 3.5 point from lower (later) cycles to earlier ones, implying that inter-iteration memory constraints are violated by this schedule.

In other words, the write to `M[i+1]` actually precedes the read of `M[i]` in the pipelined loop. That means that this loop cannot be pipelined with initiation interval 1 (or indeed 2) unless you tell the scheduler that the memory accesses of successive passes through the loop are in fact accessing different memory locations.

Such false constraints between memory operations in a pipelined loop can be removed by the bc_shell command `ignore_memory_loop_precedences`. Notice that you must be sure that the addresses are in fact disjoint: you will have made yourself responsible for this aspect of the correctness of the synthesized design.

```
bc_shell> ignore_memory_loop_precedences {op1 op2}
```

Another way that use of RAM can lead to unschedulable designs happens when the RAM used has a multicycle access cycle (e.g. a two-cycle read). In this case RAM operations are scheduled into multiple cycles. This can create surprises for a user whose design has tight latency constraints.

The RAM operations look innocuous in the original source code (they are just array references), but each RAM access takes two or more cycles, and the design cannot meet its latency constraints. The strategy for dealing with this is to ask how many cycles the memory will take, and set your timing constraints accordingly. Alternatively, consider taking some of the memory functionality out of the memory and putting it into registers (ordinary variables).

Another common problem is forgetting to declare an array variable as a memory, or misspelling the declaration. When this happens, the array of words becomes one large register; reads of individual words become bus rippers; and writes become large multiplexers. The special constraints on memory reads and writes disappear, because the whole thing is just a register; but the extra logic involved with multiplexing and ripping can swamp the rest of the circuit.

Furthermore, logic optimization time can increase to the point of impracticability, as the technology mapper attempts to optimize the mappings of thousands of flip-flops to form a single large register. For these reasons, it is often advisable to check that your memory declarations really worked the way you thought they would; this is confirmed by inspection of the scheduling reports, where the presence or absence of memory read and write operations is diagnostic. This should be done before compilation: otherwise you may wait a long time.

### Records and other structures

BC schedules memory accesses as complete and indivisible operations. Thus if a memory contains records, accessing a single field means accessing the whole record.

One important optimization strategy is to look at various ways to break up record-structured arrays in your source code. For example, you can convert an array of multifield records into multiple, separate arrays of single fields. This means that accesses to different fields are decoupled, and that it is not necessary to pack and unpack the records. BC will not do this automatically; it is up to you to understand the access patterns of an array of records and construct the array variables accordingly. A similar consideration applies to multidimensional arrays such as matrices. See Chapter 7 for an example of an optimized memory format.

## 3.7 Synthetic Components

A *synthetic* component is a component that is synthesized on the fly when needed, rather than saved in a library. Examples of synthetic components are adders, subtracters, and multipliers. Synthetic components are encapsulated in DesignWare libraries; they are the primary sharable resources (others include, e.g., I/O ports and registers). Synthetic components can be shared during allocation, as can registers and I/O ports. Other components are not shared. Thus, for example, random logic is not shared, unless you take special measures.

The DesignWare libraries provided by Synopsys and other vendors give you access to many different module definitions, and one or more implementations for each module. Thus, for example, the HDL operator '+' can be mapped to either an adder module or an adder/subtracter module; each of these might in turn have multiple implementations, each with a different realization of the carry chain. These are synthesized on the fly, because the number of combinations of technology libraries, implementation styles, input bitwidths, and timing constraints would otherwise occupy excessive disk space.

### DesignWare Developer

A common situation occurs when you want to define a new function or procedure and use it in more than one place in the source HDL. The function itself is not in the DesignWare library you already have, but you want the hardware that performs this function to be shared. Synopsys offers a product, DesignWare Developer, that lets you define your own DesignWare modules and implementations; these can have multiple implementations, can be bound to HDL operators by HDL Compiler, and so on. For example, many of our DSP-oriented customers want to use synthetic multiply-accumulate (MAC) components. Others want to encapsulate repeated hunks of random logic.

Even if you don't want it to be shared, it is sometimes very useful to be able to define a new synthetic component in order to embed control logic in the datapath. A case like this came up once when one of our customers was designing a Viterbi decoder. The decoder had many phrases of this form in the HDL:

```
      -- VHDL                // Verilog
if (cond_bit) then       if (cond_bit)
   x := data1;               x = data1;
else                     else
   x := data2;               x = data2;
end if;
```

BC normally handles this kind of branch with branches in the control FSM. That is, the value `cond_bit` becomes a primary input of the FSM, and a multiplexer is generated whose output drives a register, and the FSM includes logic to drive the controls of the multiplexer.

The problem with this is that the size of the FSM input space (i.e. the number of Boolean cubes that it can have) is exponential in the number of its inputs. So if, for example, there are twenty of the above phrases, there could be twenty input bits to the FSM, and the Boolean space of its input has a million or so minterms. This kind of thing can swamp FSM Compiler.

One way to handle this is to build a DesignWare part that is really just a mux[1]. The process whereby this is done is described in detail in Appendix A. Briefly, you define a module that captures the mux's desired behavior. You then use a function call instead of the original conditional statement; and you map the function onto the new module using the `map_to_module` pragma. This pragma tells the elaboration software that you want the DesignWare part to be used instead of, e.g., inlining the function as it stands.

The function call will then be mapped to a single CDFG node. Thus the conditional will be computed entirely inside the datapath, and so the control FSM will need no logic to steer the data.

The function will also be shared wherever the cost functions show it to be cheaper; in this case, however, sharing is not as important as the decrease in FSM size. Sharing a mux or other simple combinational block only really saves you datapath gates if its operand and result sources and sinks are already shared, so that no further muxes will be needed to share the simple function. Again, the idea here is to move conditional logic from the FSM to the datapath.

DesignWare Developer is bundled into the BC package because it is so important in such cases. The use of preserved functions, procedures, and tasks is another way to accomplish many of the same goals.

## 3.8   Preserved Functions

BC allows functions, VHDL procedures, and Verilog tasks to be used in the HDL text. In the following discussion, we will use the word *subprograms* to denote all three. In early versions of BC, subprograms were always inlined during elaboration; thus their operations and variables were shared on the same hardware resources as those of the surrounding HDL text, and in fact operations that resulted from inlining were indistinguishable from other operations.

More recent versions of BC allow the user to control subprogram inlining. By default, subprograms are still inlined; but the user can choose to prevent inlining. If a function is not inlined, it is kept as a level of hierarchy through gate-level optimization. Such a function has most of the properties of a DesignWare part.

---

[1] Actually, if it is just a mux you want, and you don't care if it can be parameterized, then function preservation (next section) is usually sufficient.

The way a user prevents inlining is with an **preserve_function** pragma inside the body of the subprogram definition.

```
-- VHDL
function func_id (...) is
-- synopsys preserve_function

// Verilog
function [7:0] func_id;
// synopsys preserve_function
```

Inlining is controlled at the level of subprogram definitions, as opposed to instantiations. Thus either all of the invocations of a subprogram are inlined, or all of them are not.

The **preserve_function** directive can be used only if the subprogram does not contain certain specifically behavioral constructs. The constructs in question are signal reads and writes; sequential DesignWare parts; clock edge statements; and rolled loops.

A subprogram that is preserved is treated as if it had become a synthetic library component. Thus it is converted into a single sharable resource, whose use is scheduled and shared just as an adder's is. It is, for most purposes, equivalent to a DesignWare part in a synthetic library.

But a preserved subprogram doesn't have some important features that are present in DesignWare parts. First, unconstrained types are not allowed, so use of subprogram preservation is restricted to subprogram calls where the interface bit widths match exactly. Second, automatic implementation selection is not possible. This may be irrelevant also; most of us don't define multiple implementations of a single function in any case. Implementation selection of DesignWare parts contained in the subprogram is handled normally. Finally, a preserved subprogram cannot be used in an RTL process.

## 3.9 Pipelined Components

Combinational logic implemented as synthetic components sometimes has an inconveniently long propagation delay. For example, multiplication operations often have a delay that is longer than the desired clock period. In this situation, you have three choices.

First, you can lengthen the clock cycle. Often this is not a real option; and sometimes it carries an unacceptable hardware cost, as the level of chaining to perform a given calculation increases (and hence sharing decreases).

Second, you can allow the 'slow' operator to take multiple clock cycles. Multicycling has a latency penalty, because all inputs to the operation must be registered so that they will remain stable during the operation.

Finally, you can pipeline the component. A pipelined component is constructed by using behavioral retiming (**optimize_registers**) to pipeline a combinational component. The combinational function $f(x)$ is split into two parts, $g$ and $h$, and a register put between the two. When each part has a combinational delay less than the clock period, we can successfully use the aggregate component as a pipeline; in each cycle, we present a new input to $h$, and a cycle later we read the result at the output of $g$.

There are pipelined components in the DesignWare libraries available from Synopsys and other vendors. Alternatively, you can construct your own using Design-Ware Developer. See Appendix A for details on constructing your own sequential DesignWare.

In recent releases of BC there is an automatic facility for pipelining components. This automatically inserts retiming register(s) in the datapath at the combinational unit to be pipelined; the result is a pipelined component that appears without further effort on the user's part.

```
bc_shell> set_pipeline_stages {op1 op2} -fixed_stages 3
```

## 3.10 Random Logic

'Random logic' in this context refers to collections of combinational logic primitives such as gates and registers, but excluding synthetic components. These can be connected together into complex collections; for the purposes of this discussion we can consider any subnetwork consisting only of simple gates to be 'random logic'.

There are four basic ways to include random logic clouds in your BC input.

1. As expressions in a behavioral process.

2. As expressions or instantiated logic outside processes, or in RTL processes.

3. As functions.

4. As encapsulated DesignWare.

It is quite reasonable and ordinary to include Boolean-valued expressions inside a behavioral process. For example, one might want to compute a sum of products:

```
x := (a and b) or (c and d) or (e and not f);
```

```
x = (a & b) | (c & d) | (e & !f);
```

The result of elaborating these statements is a gate-level netlist. The gates will not normally be shared; they form a fixed part of the datapath. The one restriction on this kind of random logic is that you cannot infer tristate logic. That is, the value Z is disallowed.

The drawback of this kind of random logic is that it is not sharable. Suppose you had to compute the same function of many data inputs at many potentially different times; then it might be worth while to create a sharable resource that would compute the function. This is achievable using either function preservation or DesignWare Developer. As mentioned in Section 3.7, this can sometimes have a profound effect on overall resource requirements.

One can also define random logic using RTL processes. Here the process will go through the HDL Compiler flow; scheduling of such processes is left to the user, and if the process is combinational the RTL process will simply be compiled. Tristates are allowed in RTL processes. Thus if your BC process needs a tristate interface, the thing to do is write a thin RTL wrapper for the BC process.

Finally, one can instantiate a netlist outside the process by instantiating gates and nets. In VHDL this is done using component and signal definitions and instances; in Verilog, one uses module and gate instantiations interconnected by `wire` variables. Again, tristate logic is legal here. BC does not do anything with logic outside the process, so it will not be changed until gate-level optimization.

Notice, however, that all communication with the world outside a behavioral process is done via VHDL signals and Verilog variables; and that the semantics of both languages require register-like behavior on the outputs of behavioral processes. Thus there is a delay between when the behavioral process writes a value and when the outside logic responds to it; this delay is not present when the logic is included in the behavioral process. Moreover, the restrictions on I/O described in the next chapter apply to all communication between behavioral processes and outside logic.

### Summary

This chapter has covered the basic aspects of HDL descriptions for behavioral synthesis, with some details that pertain strictly to Synopsys BC. You should now know that a process is the basic unit of behavioral synthesis; that the process is described in such terms that it can be made to behave indistinguishably before and after synthesis; and that this restricts the class of things you are allowed to describe to those that can also be synthesized with untransformed cycle-level timing. You should know what BC considers to be an 'operation', and what kinds of operations are scheduled; you should be aware of loop structures and of the nature of loop pipelining, and the ways in which memory and I/O operations are inferred from the source description.

You should also be coming to the realization that you need to know more about the ways in which I/O may be transformed while still preserving the 'correctness' of your design; and that 'correctness' is an elastic term in this context.

# Chapter 4

# I/O modes

This chapter describes the three I/O modes of BC. Particular attention is paid to the kinds of HDL descriptions that can be scheduled in the three modes; this is one of the most important things to learn well. You should pay particularly close attention to the cycle-fixed I/O mode, even though you may not use it often; the constraints on what you can write in cycle-fixed mode can be mapped exactly into constraints on what BC can implement. In other words, if you can't say it in cycle-fixed mode, you probably can't implement it using BC.

This chapter contains many specific rules, constraints, restrictions, and special cases that pertain to BC at the time of writing. In future releases these will almost certainly be relaxed, one at a time; the current release is much looser about some things than the one before it. So be careful: when in doubt, consult the most recent documentation. The principles and modes of thinking about and working around scheduling problems will be the same, though: you should study them even though some particulars will become obsolete.

Recall that there are two kinds of constraining mechanisms in BC. This chapter describes constraints implicit in the semantics of the HDL. HDL semantics can be interpreted or transformed in a number of ways by synthesis tools; for example, logic synthesis ignores most of the HDL semantics having to do with propagation delays. In behavioral synthesis we are in a position to take many more liberties with the HDL semantics. The three I/O modes of BC represent three different abstractions or interpretations of the HDL semantics.

Underlying the definition of BC's I/O modes is the question of what it means for two designs (or HDL texts) to be equivalent. This question has broad implications for synthesis in general, but scheduling makes it particularly interesting and important. The two designs being compared in this context are the HDL text before synthesis begins, and the design after scheduling. In this context the term 'timing' denotes the *state* in which an I/O operation occurs; we will not consider propagation and setup delays of less than one clock cycle, because these are difficult or impossible to synthesize accurately.

A very strict definition of equivalence is that two designs are considered equivalent if and only if they perform the same operations on their input data at the same times. That definition would rule out scheduling altogether. A somewhat looser definition would say that two designs are equivalent if and only if their I/O behavior was always the same. Such a 'black box' definition in effect says that the internal operations of a black box could be radically different, but the two circuits or HDL texts would still be equivalent if they could not be distinguished from one another merely by observing their inputs and outputs. BC's cycle-fixed I/O mode works that way. A still looser definition, commonly assumed in the literature on scheduling, allows I/O operations to be freely shifted in time; that definition is roughly the same as BC's free-floating mode. Both interpretations are 'correct' in some sense; both interpretations have their uses.

In defining the I/O modes of BC we had to balance predictability and intuitive fit against the size of the solution space for scheduling and the amount of detail the user is forced to supply. Clearly, a very restrictive interpretation (such as the cycle-fixed mode) is the easiest to predict and understand; but under some circumstances it may represent a needless restriction of the solution space that BC is permitted to explore, and hence a needless limitation of the achievable quality of results. A restrictive interpretation also places the greatest burden on the user, in that the HDL timing must be synthesizable, and therefore that the user must know what BC can and cannot achieve.

On the other hand, a looser interpretation of the HDL semantics gives both BC and the user a great deal of freedom, but can result in radical changes of I/O behavior in the design. Whether and to what degree you can tolerate changed I/O behavior seems to depend on the problem and on where you are in the design process. For example, in the very early stages of design I/O protocols might not be defined, and you might just want to find out what is achievable given a particular algorithm; in the later stages protocols might be more rigid.

The three I/O modes of BC represent different definitions of what it is to be 'equivalent'. Each I/O mode is a different way of interpreting and transforming the HDL semantics; each has advantages and disadvantages. The three modes are, in outline, the *cycle-fixed* mode, in which I/O operations are not free to be rescheduled; the *superstate-fixed* mode, in which I/O timing can be stretched but not permuted; and the *free-floating* mode, in which I/O operations are free to float.

A graphical representation of the three BC I/O modes is shown in Fig. 4.1. At the top is the clock; it is common to all four of the other timing diagrams. Immediately below it is the original source HDL timing of an example circuit. The next timing diagram down is for the cycle-fixed mode; notice that it is the same as that of the source.

Cycle-fixed mode preserves the original I/O timing of the source HDL exactly. It has the great advantage that it does not transform I/O timing at all, with the one exception of resets. This means that a communication protocol that works for the input HDL text will also work for the synthesized design; that your test bench will work without modifications; and that the HDL description has absolute control over when I/O operations occur. It enforces a strict discipline in that you must specify achievable I/O timings. For example, you might read in a sample in one cycle, perform some computation, and write out a result in the following cycle. If

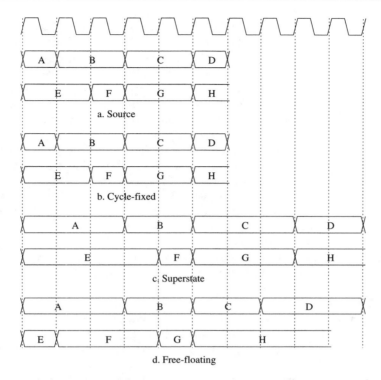

**Figure 4.1.** Original HDL timing and results of three I/O modes

the computation takes too long, no schedule satisfying the constraints exists and BC will exit with an error message. Fixed mode makes you responsible for defining I/O timings that can lead to achievable (and high-quality) implementations.

Below the cycle-fixed timing in Fig. 4.1 is a superstate-fixed timing of the same circuit, with one choice of latencies; notice that the relative ordering of the signal transitions is not changed, but the numbers of clock cycles between signal transitions has, in some cases, been stretched.

The superstate-fixed mode has the property that the design's I/O timing after synthesis can only be distinguished from that of the input HDL by counting numbers of clock edges. This means that one test bench can be made to function with both the pre- and post-synthesis designs, if only it is written to be independent of numbers of clock edges. Alternatively, you can think of superstate mode as 'stretching' the amount of time between I/O transitions, without permuting any two transitions.

Notice that the word 'transitions' in this context refers both to input and output transitions. What that means for an output write is obvious: the stabilizing register associated with the output port has its load enable active during the states corresponding to the csteps in which I/O writes are scheduled on the port.

The proper way to think about input port transitions is to think of the times a read operation can safely occur on the port. Clearly, it is hazardous to allow inputs to transition at such times that a read operation could be scheduled either before or after the transition. Thus the normal practice is to ensure that inputs will not transition except at times defined by the property that all read operations are constrained to be scheduled either before or after the transition time. In the superstate mode one such set of times are defined by the clock edge statements of the source HDL.

The advantage of superstate mode is that the latency of your schedule can be varied by a single bc_shell command; no source HDL changes need be made. Another advantage is that BC can fit the latency to the computations that are being performed; the possibility of unsatisfiability because of a simple lack of time is greatly reduced. There are two disadvantages of superstate mode. First, it is more restrictive than either fixed or floating modes. That is, there is a class of HDL texts that is legal in the other two modes, but that is illegal in superstate mode. Second, in superstate mode you must be careful either that read operations are tightly constrained, or that input signals always transition on boundaries that no read operation can cross during scheduling.

At the bottom of Fig. 4.1 is the free-floating timing; here, both stretching and transition permutation has occurred. The free-floating mode has the properties that reads on a single port can be permuted, and reads and writes on different ports can be permuted. This gives the scheduler the most freedom to optimize; it also can cause the I/O timing to be radically changed across synthesis. This is a big advantage in the initial exploratory phases of design, and when the user has timing constraints that cannot be readily expressed in the input HDL; but it is also the most slippery of the I/O modes.

## 4.1   Cycle-Fixed Mode

Cycle-fixed mode (henceforth just 'fixed' mode) is a scheduling mode that does not transform the I/O timing of the HDL, with a single minor exception having to do with reset functionality.

It is important to understand fixed mode thoroughly, even if you intend to use one of the other modes. The reason it is important is that the shortest latencies achievable by BC in any mode have exact fixed-mode HDL counterparts. In other words, if you cannot achieve a particular timing diagram in fixed mode, you cannot achieve it in any mode; what BC will let you do in fixed mode is as good as you can get using BC.

So, for example, if fixed mode forces a control step boundary between evaluation of a condition and a state branch that depends on the condition, that boundary will also be present in the other modes, even though there is no need to put in an explicit clock edge. In the other modes, BC will just add the extra control step automatically, if it can; and exit if it cannot. This means also that you can't get around the fixed-mode constraints by using manual constraints and another mode;

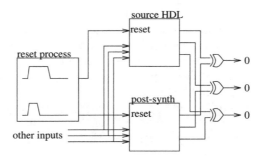

**Figure 4.2.** Reset pulse to the source is extended 1 cycle

BC will just reject the design if you try this. You can always find a fixed-mode description that matches any timing you can achieve using another mode.

The definition of fixed mode has two important consequences. First, if the source HDL process can talk to its environment (i.e. its protocols are correct) then the synthesized process will also have correct protocols. Second, you must write the source in such a way that it can be synthesized by BC without BC having to add or delete any clock cycles. This means that in order to get the most out of BC in fixed mode, and therefore also in the other modes, you have to have an understanding of how BC schedules and allocates hardware.

The single exception to the identity of behavior in fixed mode is concerned with resets. When simulating the source text, the reset pulse need not be present at all: a process always begins at its beginning when simulation starts. In the synthesized design this is not the case: at least one cycle of reset, and sometimes more, is needed to properly start up the control FSM.

Furthermore, BC always registers a process's output signals. Registering the outputs results in a single cycle of skew between simulation of the source and the synthesized design. This skew results in a trivial mismatch of behavior when simulating the two designs side-by-side. The easy way to take care of this mismatch is by providing two reset pulses in your test bench, one with a duration of $k$ cycles, where $k$ is the minimum necessary duration to reset the synthesized circuit; and one that produces a pulse of duration $k + 1$, which will be delivered to the source design and will keep it quiescent for the extra cycle. This is shown in Fig. 4.2.

Other than the reset pulse, if your test bench is properly constructed the pre- and post-synthesis designs should simulate in exactly the same way. One caveat: you should be sure that test bench signals other than the clock don't transition exactly on the clock edge; this can cause races in zero-delay simulations. If you are doing equivalence testing between the pre- and post-synthesis designs, you should test the outputs only on clock edges; otherwise you will see setup and hold time variations. This is most easily done using VHDL **assert** and Verilog **@(posedge ...)** **and $display** statements.

A cautionary note having to do with resets is appropriate here. When a reset occurs the state of the circuit is to a large extent[1] preserved; so, in particular, if you had a variable that was read before it was written in your source HDL, the read would see the last value that was stored in the variable before the reset pulse occurred. Scheduling changes the times when operations occur; so the result of reading an uninitialized variable can be expected to differ in the pre- and post-synthesis designs. We didn't think this was a very important difference, because we don't think that many designs will make use of the data in uninitialized variables, so we made such use illegal. If you really want this behavior, you can get it by using a signal instead of an internal variable; but your troubles will only have begun.

## 4.1.1 Rules for Fixed Mode

The following rules are consequences of the statement that you must specify only I/O timings that can be synthesized and of the kinds of schedules and circuits that BC is capable of constructing. This should be borne in mind at all times; when in doubt, just ask yourself whether BC could synthesize what you are writing. If not, or if BC would have to take heroic measures to synthesize it, you will probably end up rewriting your source.

### Latency of straightline code

If BC detects a situation wherein a series of operations are connected by data or by strong precedence arcs (see section 2.2.1), and the operations don't fit in the allocated time, then it will report unsatisfiability and exit. This calculation takes chaining, sequential, and multicycle operations into account. It also takes manual constraints into account.

The one thing that you might not expect is that multicycle operations can only be chained with output write operations. That is, no single-cycle operation ever drives a multicycle operation, and no single-cycle operation except an output write is ever driven by a multicycle operation, unless there is an intervening register (which in turn implies an intervening cstep boundary). The reason for this restriction has to do with the way timing constraints are constructed for logic synthesis: briefly, the timing paths for multicycle operations must be disjoint from the timing paths for single-cycle operations. From the point of view of logic synthesis, an output write is just a register write: so it can be excused this constraint.

A common source of problems is failure to appreciate the effects of multicycle and sequential operations. These increase latency, but that isn't always immediately obvious from inspection of the source HDL. This is particularly sneaky in the case of memory operations, which just look like array references, but which in fact can take two or more cycles, and which have implicit ordering constraints because BC doesn't do address equivalence inferencing (see section 3.6).

---

[1] Consult the documentation on the variable `reset_clears_all_bc_registers` for an alternative.

## Loops

In fixed mode loop boundaries are not free to be rescheduled. The reason this is so is that moving loop boundaries will almost always cause changes in the numbers of clock edges between pairs of externally visible events. Consider, for example, a pair of writes occurring before and after a loop. If the loop beginning were to be moved one cycle later, then each iteration of the loop would take one cycle less; so the total time in the loop, and hence between the I/O operations, would decrease.

The only case where moving loop boundaries would not cause changes in visible timing would be if every boundary associated with the loop moved together in a rigid way, and clock edges moved over the loop to compensate. At the time of writing BC doesn't support this case.

In BC a loop is mapped to two or more csteps, and thence to a collection of one or more states and one or more transitions arranged to form a cycle. The loop test (e.g. the conditional part of a **while** statement) is contained somewhere inside the cycle. The result of this is that the loop must be entered in order to perform the test; that is, there is no transition that goes 'past' the loop.

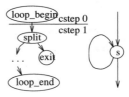

The most important result of this state graph structure is that a loop always takes at least one state, even if it is a while loop whose condition is false on entry; the fixed-mode HDL structure reflects this. The form the 'extra' state takes is an extra clock edge between the test of the condition that exits the loop and the next I/O operation, or any other operation that is fixed in time. The extra state gives the FSM time to figure out what is happening. The need for the extra cycle is most readily apparent in fixed I/O mode, but its effects are apparent in the other modes as well: e.g. when constraining operations that fall on different sides of a loop.

The following example is illegal because there is no clock edge between the test of **ready** and the succeeding output write. If we inserted a clock edge statement immediately following the end of the loop, it becomes legal.

```
-- VHDL                          // Verilog
while (not ready) loop           while ( ! ready ) begin
   wait until clk'event             @(posedge clk);
            and clk = '1';
end loop;                        end;
-- Illegal! no wait              // Illegal: no @edge
dataout <= data;                 dataout <= data;
```

Internally, the write immediately following the loop has a constraint, expressed as a template, forcing the write and the loop beginning to fall in the same cstep (count the clock edges if **ready** is true). This creates a contradiction, because the exit can't be scheduled later than the write. Look at the single-cycle loop state diagram and ask yourself where the write could go: on the upper transition, where the write would be 'before' the loop; on the loopback, where the write would occur only if the loop were executed; or on the exit transition, where the write would occur one state too late if **ready** was true.

The most accurate way to think about **while** loops is as unbounded loops with conditional exits; there must be a clock edge between the decision to exit and the exit itself. Internally this is represented in BC as a strong precedence of length one between the split that represents the conditional and the exit node that gets you out of the loop. Rolled **for** loops are internally represented the same way as while loops; the test and iterator are automatically generated.

```
-- VHDL                              // Verilog
while_loop: loop                     begin: while_loop
                                        forever begin
   if (ready) then                        if (ready) begin
      wait until clk'event                   @(posedge clk);
            and clk = '1';
      exit while_loop;                       disable while_loop;
   end if;                                end
   wait until clk'event                   @(posedge clk);
            and clk = '1';
end loop;                               end
                                      end
dataout <= data;                     dataout <= data;
```

Where loops are tightly nested we have restrictions that stem from the fact that BC requires a cstep between a conditional and a state transition branch that is based on the conditional. This fragment is illegal for two reasons.

```
-- VHDL                          // Verilog
while (not done) loop            while ( ! done ) begin
   -- position A                    // position A
   while (not ready) loop           while ( ! ready ) begin
      wait until clk'event             @(posedge clk);
            and clk = '1';
   end loop;                        end
   -- position B                    // position B
end loop;                        end
```

The first problem is where control is transferred from the outer to the inner loop; this is illegal because as it is written, both conditions must be tested in one state. Thus a clock edge is needed at position 'A'.

The second problem happens when the inner loop is exited. The loop test of the inner loop takes one clock tick, which would also be the case if the inner loop were not nested. We therefore need a second clock edge at position 'B'.

Another case where there isn't enough time to figure out what to do occurs when loops are neither nested nor separated by a clock edge.

```
-- VHDL                              // Verilog
while (not done) loop                while ( ! done ) begin
    wait until clk'event                 @(posedge clk);
              and clk = '1';
end loop;                            end
-- illegal: no wait here             // illegal: no posedge
while (not ready) loop               while ( ! ready ) begin
    wait until clk'event                 @(posedge clk);
              and clk = '1';
end loop;                            end
```

Here the problem is between the loop test of the first loop and the loop test of the second; both would end up on the same state transition unless we put in a clock edge between the two loops.

You should also remember that the time to evaluate the conditional is added in to the 'extra' delay. If the conditional is just a bit test, as above, it usually isn't a problem; but suppose the loop had a complicated condition that couldn't be computed in the csteps before the test:

```
while ( x * inport1 < y - inport2 ) ...
```

Here two of the operands are actually I/O reads, which are locked into the first state of the loop; after the reads, the multiplications take place, and then the comparison. This takes time in the circuit, if not in the simulator: the result is 'extra' cycles of latency, which must be modeled as additional edge statements inside and after the end of the loop.

## 4.2  Superstate-Fixed Mode

In superstate-fixed (henceforth 'superstate') mode the idea was to preserve the ordering of I/O operations but not necessarily the number of clock edges between I/O operations. This means that the latency of the design can be varied without changing the HDL source, through the use of commands such as **set_cycles** in shell scripts.

```
bc_shell> pipeline_loop main_loop -latency 12 -initiation 3
```

The reason it is called 'superstate' is that we regard the interval of time between two clock edges of the source HDL as a collection of states (a superstate) arranged in a branch-free chain of states and transitions. BC can expand the interval between two clock edges by adding new clock edges (states) to the superstate; the I/O sequence is preserved by forcing all writes to remain in the last state of the superstate.

**Equivalence**

I/O equivalence is maintained within the superstate abstraction by the following two rules:

    1. Any I/O write in a superstate will take place in the last cycle of the superstate.

    2. An I/O read in a superstate can take place in any cycle of the superstate.

The fact that all I/O writes of a superstate take place in the last state of the superstate implies that two writes falling between the same two clock edge statements of the source HDL must result in simultaneous output transitions in both the source and the synthesized design.

The ability of I/O reads to float freely between the boundaries of the superstate means that input data must be held stable within a superstate. This is analogous to the assumption that inputs are stable throughout the clock cycle. That is, BC assumes that external logic registers all inputs to the BC process.

Having the I/O reads float means that you must take care that processes driving BC inputs do not transition their outputs during superstates when the inputs are read. Imagine, for example, a situation in which two reads of the same port fall in the same superstate. In the original HDL, these reads would always see the same data. But in the superstate-mode synthesized design, the two reads might easily be in different states of the same superstate. Thus if the input transitioned between the first read and the second, your circuit would see two different values, with whatever results that might cause.

In addition, the ability of superstates to stretch means that your I/O protocols must be capable of dealing with inserted delays. If you use handshaking (as is usually the case if you have data-driven branching) this is not normally a problem. Then it is sufficient to ensure that external logic has some way of knowing when the important superstate boundaries have occurred, and that the external logic holds its data stable during superstates when that data might be read.

## 4.2.1   Rules for superstate mode

The nature of superstate mode gives rise to five rules that must be enforced if behavioral anomalies are not to be created. The rules don't pertain to fixed I/O mode, because in fixed mode no new clock edges can be added, and I/O operations cannot be moved past clock edges; hence the question of I/O operations moving from one state to another is not an issue in fixed I/O mode. But in superstate mode I/O operations can move over the clock edges that BC inserts; in order to prevent violations of the definition of equivalence, we had to restrict what the HDL text can say.

Consider, for example, the loop of the next example, which contains an I/O write in a position that is illegal in superstate mode.

```
-- VHDL                              // Verilog
while (not ready) loop              while (!ready) begin
   tmp := inport; -- I/O read          tmp = inport; // read
   wait until clk'event -- clock edge 1   @(posedge clk);
            and clk = '1';
   wait until clk'event -- clock edge 2   @(posedge clk);
            and clk = '1';
   outport <= data; -- illegal         outp <= data; // illegal
end loop;                           end
```

The reason the I/O write is illegal is that the superstate beginning with clock edge 2 can end at clock edge 1. That is, in execution order the two clock edges are adjacent (and hence form superstates) in both the ordering edge 1 – edge 2 and the ordering edge 2 – edge 1. Now consider the superstate generated by the ordering 2 – 1. The write operation must migrate to the end of this superstate. Thus it must be scheduled after or cotemporaneously with the read that precedes clock edge 1. But on entering the loop the write would not take place at all; so there is an internal contradiction, which cannot be statically resolved.

We call a superstate that begins somewhere in the body of the loop and ends at the first clock edge of the loop a *continuing superstate*. The last superstate of a loop is a continuing superstate; so is a superstate containing a VHDL **next** or a Verilog **disable**.

The reason the contradiction occurred was that we had a continuing superstate that had an I/O write, and the first superstate of the loop had an I/O operation. Thus we have rule 1 of superstate mode.

*If the first superstate of a loop contains any I/O, no superstate containing a loop continue may contain an I/O write.*

The second rule of superstate mode is needed to prevent a contradiction in situations like the one shown here.

```
   -- VHDL                           // Verilog
   thisport <= something;            thisport <= something;
   -- illegal: no wait               // illegal: no @edge
   loop                              forever begin
       thatport <= anything;             thatport <= anything;
       wait until clk'event              @(posedge clk);
                and clk = '1';
   end loop;                         end
```

Here the contradiction arises because the first superstate of the loop begins outside the loop. The write that occurs before the loop migrates to the end of the superstate, i.e. into the loop, where it will conflict with the loop's write. Hence the second rule of superstate mode.

*There must be a clock edge between a write operation and the beginning of a loop whose first superstate contains a write operation.*

The third rule of superstate mode prevents scheduling ambiguities in situations such as that of the next example, in which the contradiction arises from the possibility that the loop may not be executed at all.

```
-- VHDL                                    // Verilog
thisport <= something;                     thisport <= something;
-- illegal: no wait                        // illegal: no @edge
while (strobe) loop                        while (strobe) begin
   wait until clk'event and clk = '1';        @(posedge clk);
end loop;                                  end
storage := thatport;  -- I/O read          storage = thatport;
```

If the loop is never executed, the write preceding the loop should, by the definition of superstate mode, migrate over the loop and to the end of the superstate following the loop. If the loop is executed, on the other hand, then the write will occur in the first superstate of the loop. This is an ambiguous or dynamic schedule. The same kind of problem occurs with any conditional clock edge. Conditional clock edges create conditional superstate boundaries.

*A write can never precede a conditional superstate boundary if any I/O operation succeeds the boundary.*

The fourth rule of superstate mode prevents a write operation from 'escaping' out the back of a loop. Suppose we have a loop with a conditional exit, and an I/O operation follows the loop with no intervening clock edge. Then *no I/O write can occur between the exit and the last clock edge before the exit.* If there were such a write, then it would migrate past the exit and out of the loop, and hence it would conflict with the I/O operation following the loop.

```
-- VHDL                                    // Verilog
                                           begin: busy
busy: while (strobe) loop                     while (strobe) begin
   wait until clk'event                          @(posedge clk);
              and clk = '1';
   thisport <= something;                         thisport <= something;
              -- illegal!                                 // illegal!
   if (interrupt) then                        if (interrupt)
      exit busy;                                  disable busy;
   end if;
end loop;                                     end
                                           end
storage := thatport;                       storage = thatport;
```

## 4.3  Free-Floating Mode

In free-floating mode I/O operations are free to float with respect to one another. The operations taking place on a single port are partially ordered: the reads can be permuted. Free-floating scheduling preserves this partial ordering. By contrast, the orderings of I/O operations on different ports are ignored. I/O operations do not enter or leave loops. Data precedences are respected as well, so for example an I/O write cannot take place before its data is computed. With those exceptions, I/O operations are scheduled whenever they save the most hardware and manual constraints permit.

Free-floating mode also permits both the insertion of new clock edges and the deletion of clock edges specified in the source HDL. The ability of the scheduler to delete clock edges is unique to free-floating mode; superstate mode can only add edges, and fixed mode can neither add nor delete edges.

It is very important to set your constraints carefully when using free-floating mode. For example, a very easy error to make in building free-floating designs is failure to constrain I/O operations and their synchronizing signals to float together. Consider, for example, the case where a bus is synchronized by a data-ready strobe. The bus transfer will float in time to the step where it uses the least resources and wastes the least latency; this is determined by its data dependencies, manual constraints, and so on. But the strobe depends on essentially nothing; its data is an otherwise unconstrained constant. As such, it tends to float to the beginning of the schedule: thus eliminating its value as a strobe. So you must constrain the data and the strobe to be concurrent using set_cycles; then they will float together and your synchronization will work.

The best way to use free-floating mode seems to be to define your I/O in tightly constrained blocks, which are then free to float as blocks, with looser constraints tying the blocks together. Another way to use free-floating mode is as a way of exploring the various latency/hardware issues in the context of a minimal set of global latency constraints; but the latency/hardware tradeoff curves so generated should be used with some caution, because the addition of real I/O constraints may result in increases in latency and gate count.

On the next page is an example of an HDL source file and a corresponding BC script, in which a design is constructed using free-floating mode. This example is given in Verilog in order to conserve space.

The design consists of a loop and a case statement; the case statement contains other loops. The first loop (**sync**) is a busy wait loop, exiting only when either or both of two one-bit input ports **m1** and **m2** goes high.

The case statement switches between four modes, as determined by the contents of the two ports when the case is switched. The modes in question create a five-cycle delay, a ten-cycle delay, sum the contents of the input port **din** twice, or zero out the accumulator **tmp** respectively. Then the contents of the accumulator are written out to the output port **dout** and the cycle is repeated.

```verilog
module free (m1, m2, din, dout, clk, reset);
   input m1, m2, clk, reset;
   input [7:0] din;
   output [7:0] dout;
   reg [7:0] dout;
   always begin: main
      reg [7:0] i, tmp;
      tmp = 8'h00;
      @(posedge clk);
      begin: sync
         forever begin
            if (m1) disable sync;
            if (m2) disable sync;
            @(posedge clk);
         end
      end
      case ({m1, m2})
         2'b00:
            for (i = 0; i < 5; i = i + 1) begin: wait_5
            /* synopsys resource foo: dont_unroll = "wait_5"; */
               @(posedge clk);
            end
         2'b01:
            for (i = 0; i < 10; i = i + 1) begin: wait_10
            /* synopsys resource foo: dont_unroll = "wait_10"; */
               @(posedge clk);
            end
         2'b10:
            for (i = 0; i < 2; i = i + 1) begin: inc_1
            /* synopsys resource foo: dont_unroll = "inc_1"; */
               tmp = tmp + din;
               @(posedge clk);
            end
         2'b11:
            tmp = 8'h00;
      endcase
      dout <= tmp;
   end
endmodule
```

The script that synthesizes this example is shown below. The **set_cycles** command used here locks its arguments into a rigid timing relationship, so that there will be a specific number of csteps between the occurrence of the first and the occurrence of the second.

```
analyze -f verilog free.v
elaborate -s free
create_clock clk -p 20

set_cycles 1 -from_begin main/sync -to_end main/sync
set_cycles 5 -from_begin main/wait_5 -to_end main/wait_5
set_cycles 10 -from_begin main/wait_10 -to_end main/wait_10
set_cycles 2 -from_begin main/inc_1 -to_end main/inc_1

set_cycles 1 -from_end main/sync -to main/m1_17
set_cycles 0 -from main/m1_17 -to main/m2_17

set_cycles 2 -from_end main/sync -to_begin main/wait_5
set_cycles 0 -from_begin main/wait_5 -to_begin main/wait_10
set_cycles 0 -from_begin main/wait_5 -to_begin main/inc_1

schedule -io free
```

Notice the explicit timing constraints in this example. These are required in order to preserve the ordering of I/O operations in free-floating mode; otherwise, for example, the read operations of **m1** and **m2** would all float up to the first cycle of the schedule and be rendered meaningless. The constraints can be broken down into three groups: four constraints are used to keep the loop durations at the right numbers of cycles; two constraints hold the first reads of **m1** and **m2** in the second cstep of the **sync** loop; and three constraints hold the beginnings of the loops **wait_5**, **wait_10**, and **inc_1** two csteps after the end of the **sync** loop.

This is reasonably typical of what you have to do when using free-floating mode. Clever use of loop boundaries as reference points to which other operations can be attached, as shown here, results in compact and intuitively clear constraint sets. Where I/O interfaces are more complicated, as e.g. the simultaneous reads of **m1** and **m2**, it is usually easiest to tie the operations to a single master operation, then constrain the master operation to a loop boundary or other clearly definable reference point.

## 4.4   Control Unit Registers

Recall that the restrictions on cycle-fixed mode are imposed by the class of designs that BC is capable of synthesizing. What BC can do, in terms of latency around conditionals, is strongly affected by whether the inputs and/or outputs of the control unit are registered.

The way control unit inputs and outputs are registered is controlled by the bc_shell command **register_control**, which allows you to force registers on the inputs and/or the outputs of the control FSM. These registers create latency between a condition and datapath actions or state branches that depend on the condition.

```
bc_shell> register_control -inputs -outputs
```

To see why, suppose that all inputs of the control FSM are registered. This means that any status information computed in cstep $i$ will not be seen by the FSM until cstep $i + 1$; in that cstep the FSM can send an output to the datapath or take a state branch in response to the status bit. For ordinary logic and synthetic operations this is not so much of a problem, because we can perform the operation anyway and then throw away the result if the condition turns out to be false.

If, on the other hand, the dependent operation is an I/O or memory write, or a state branch (e.g. a loop exit) then the FSM must simply wait until it knows which way to go. This is a *latency penalty*: between the computation of the condition and the resulting action there must be at least one cstep boundary.

The same consideration applies when the control unit outputs are registered. Here the control unit can see the status bit in the cstep in which it is computed; but the control unit can only commit to a control vector that will become visible to the datapath in the following cycle.

If both the inputs and outputs of the control unit are registered, the latency penalty is two, because the FSM cannot see the status until the step after the status is generated, and its own response to the status cannot be seen by the datapath until the step after the FSM sees the status.

In internal terms, registering the control unit inputs means that a strong precedence of length one is imposed between the node that computes a condition and the split that consumes the condition; registering the control output means a precedence between the split and anything that depends upon it. The error messages issued when no schedule exists reflect this strong precedence when it is present; so it is important to remember this internal detail when deciphering the error reports.

Now you may be wondering why an otherwise intelligent user would ever want to register the FSM at all. The answer to this question is that the clock cycle time may be reduced by pipelining the FSM and the datapath. If the critical path for logic synthesis (i.e. the critical path that determines the minimum clock period) passes through the FSM, then this may be the right thing to do. If, on the other hand, the critical latency (i.e. the number of cycles between a particular pair of operations) is the important consideration, you might want to chain the control and the datapath; that is, you might not want to insert the extra register delay or delays. Which way you go on this depends on whether the latency or the propagation delay is more important.

Generally speaking, registered control inputs are cheaper than registered control outputs, because the FSM almost always has fewer inputs than outputs.

But control FSM input (status) registers are first-class datapath registers, and all such registers must be built with load enable pins. This means that they can't be retimed very well, because of the small combinational loop from the Q to the D pin of a load-enable register.

On the other hand, control unit output registers are nearly always built without enable inputs, which means that they can be retimed with much better results. So if you intend to do retiming, and the extra gate count is acceptable, you should probably register the control outputs. Note also that internal FSM state registers are also retimable; so you may get the benefit of pipelining without registering both the FSM inputs and outputs, and thus incurring only a single-cycle latency penalty.

## Summary

This chapter describes the three modes in which Verilog and VHDL input can be interpreted for scheduling. These modes are the cycle-fixed, superstate-fixed, and free-floating modes. You should now understand the three modes and what they mean, in two senses. First, you should understand that the choice of I/O mode has an effect on the way you will write and constrain your source; and second, you should understand that the choice will have an effect on the I/O timing of the synthesized design: in two of the modes the timing will be changed by synthesis.

# Chapter 5

# Explicit Directives and Constraints

This chapter describes methods for constraining scheduling and allocation. The previous chapters covered implicit scheduling constraints, i.e. those that are implicit in the HDL text and the choice of I/O mode. In this chapter we will turn to explicit constraints. First, however, let's look at some argument naming techniques.

## 5.1  Labeling

A common factor in explicitly constraining operations is the need to name them. That is, you need to be able to specify a particular operation so that BC knows where to attach the constraint. You have a choice in this: you can make up your own name and attach it to the operation, or you can find out the name given to the operation by default.

The default naming scheme for operations in BC gives a path down into the HDL hierarchy, and then adds a line number to the end. The hierarchy is that used by scheduling; that is, loops (and sometimes subprogram calls) are levels of hierarchy. Consider, for example, the following fragment.

```
    -- VHDL                    // Verilog
  first: process            always begin: first
     outer: loop               forever begin: outer
        inner: loop               forever begin: inner
           x := a + b;               x := a + b;
        end loop;                 end
     end loop;                 end
  end process;              end
```

Let us suppose that the addition falls on line 42 of the source code in both cases. Then the default name of the addition would be "first/outer/inner/add_42" in both cases; the path being similar to a directory path in DOS or Unix, denoting a root process named **first**, and working its way down to the addition operation on

74

line 42. If there were two additions on the same line, they would get names ending in the strings "add_42_1" and "add_42_2"; other types of operations are named appropriately, e.g. mul, as you would expect. If we had not labeled the loops of the design (loop labels are optional in both languages) then suitable names (e.g. L0) would be made up for the loops. In an unrolled loop names are generated by concatenating the default name with the iteration variable and its value when the operation was copied. Hence, e.g., you might see names like "first/inner/outer/add_42_i_3".

The simplest way to find out what a particular operation is named is to use the bc_shell find command to list all of the cells in the design after elaboration. The result is a list of all of the loop and operation names; you can then use a text editor to be sure that the names you are using indeed denote the operations you mean.

```
bc_shell> find -hier cell > names.txt
```

The main drawback of this naming scheme is that you will want to write scripts around the operation names you find. This itself is not a problem; but what if you add a line to your source code? All of the subsequent line numbers will change, and your script will have to be corrected.

The second labeling scheme solves the line number problem by labeling a specific operation with a user-specified name. This is done by the use of a pragma:

```
   -- VHDL
 first: process
    outer: loop
       inner: loop
          x := a + b; -- synopsys label opname
       end loop;
    end loop;
 end process;

   // Verilog
 always begin: first
    forever begin: outer
       forever begin: inner
          x = a + b; // synopsys label opname
       end
    end
 end
```

The name of the operation will now be "first/outer/inner/opname", and it will not change when you add lines to the source HDL.

The labeling scheme shown above has limitations. First, if there is more than one operation on the line, it isn't always obvious which operation will get the label; second, if a statement occurs on multiple lines, the last line won't get labeled unless you put the trailing semicolon on a line by itself; and third, it doesn't work for I/O and memory operations, or for loop boundaries.

Recent versions of BC support an improved naming scheme that labels lines; the line label is then used to generate names for the syntactic constructs on the line in left-to-right order. Thus, for example,

```
first: process
    outer: loop
        inner: loop
            x := a + b; -- synopsys line_label thisline
        end loop;
    end loop;
end process;

always begin: first
    forever begin: outer
        forever begin: inner
            x = a + b; // synopsys line_label thisline
        end
    end
end
```

The name of the addition now is "`first/outer/inner/add_thisline`". In addition, the assignment to **x** is also a labeled operation; and I/O reads and writes, as well as memory accesses, are labeled by this construct as well. Notice that the line number is not used; thus you still have line number independence. If there are multiple operations of the same type on a labeled line, the labels are generated in left-to-right order.

## 5.2  Scheduling Constraints

Explicit scheduling constraints allow us to force the schedule to put operations into particular csteps, or to control the number of csteps between two operations.

The **preschedule** command is used to force the named operation into a particular cstep. In a hierarchical design the cstep will be the cstep of the nearest surrounding hierarchical context, with the beginning of the context set to zero; it is done this way because the schedule is constructed bottom-up. This one puts the operation **sub_107** into the fifth step of the loop **main**.

                    bc_shell> preschedule p2/res_loop/main/sub_107 4

The command **set_cycles** is used to set the number of csteps between two operations or loop boundaries. It takes three arguments: a cstep count, a leading operation, and a trailing operation. The trailing operation will lag the leading operation by the number of csteps given. For example, **set_cycles 3 -from op1 -to op2** forces **op2** to trail **op1** by exactly three csteps. The flag **-from** is a member of a family of three flags.

1. **from** ties the constraint to the beginning of the leading operation.

2. **from_begin** behaves the same way as **-from**; it is included so that the meaning of script commands can be made clearer.

3. **from_end** ties the constraint to the end of the leading operation.

For a single-cycle operation the three flags are equivalent. The three flags **-to**, **to_begin**, and **to_end** are exactly analogous to the **from** family, except in that they apply to the trailing operation.

The **from** and **to** options allow you to attach a constraint to either the beginning or the end of a multicycle operation, a sequential operation, a loop, or a subprogram call. For example, we might want to constrain the time from the beginning to the end of a loop to be three:

```
bc_shell> set_cycles 3 -from_begin loop4 -to_end loop4
```

Two other commands, **set_min_cycles** and **set_max_cycles**, are similar in function and arguments to **set_cycles**, except that they do not describe a single number, but the minimum and maximum limits respectively. This example forces the time between the end of loop1 and the beginning of loop2 to be at least five cycles.

```
bc_shell> set_min_cycles 5 -from_end loop1 -to_begin loop2
```

The commands **chain_operations** and **dont_chain_operations** are used to either force chains or break them. They are equivalent to **set_cycles 0** and **set_min_cycles 1** respectively.

The command **remove_scheduling_constraints** is used to remove all explicit scheduling constraints from a design or process. It has no effect on implicit constraints; and it cannot be controlled on a constraint-by-constraint basis.

The **set_common_resource** command is used to restrict the number of components used to implement a set of operations. The command takes as arguments the set of operations in question and the number of processors allowed; BC will then attempt to schedule the design in such a way that the named operations can all be bound to that many processors.

```
bc_shell> set_common_resource op1 op2 op3 -min_count 2
```

If a schedule cannot be constructed within your timing constraints, whether implicit or explicit, BC will add more resources to the resource set. If this happens and you are really determined to get the number of resources down, you must relax your timing constraints and/or change the HDL. If scheduling finds a solution within the resource grouping constraint, then allocation will also respect your grouping of the operations.

## 5.3  Options

Behavioral Compiler offers the user a number of options that control various aspects of the implementation it generates. This section describes most of the options available at the time of writing.

## Shell variables

The variable bc_enable_chaining, when set to false, is used to globally turn off chaining of synthetic operations. It does not affect random logic. It should be used only in situations where chaining is affecting the clock period or resource requirements adversely and the use of the more specific dont_chain and/or set_min_cycles would be too verbose. The default setting is true.

The variable bc_enable_multi_cycle is used in situations where scheduling is creating a multicycle operation, but you don't want the operation to be multicycle. For example, you might know that you can build or buy an implementation of the module in question that will meet single-cycle timing. In that case you can avoid the latency penalty associated with registering the inputs of the multicycle operation. The values bc_enable_multi_cycle can take are the strings true and false; true is the default.

The variable bc_enable_speculative_execution allows operations without side effects (e.g. arithmetic operations, as opposed to I/O and memory operations) to precede the evaluation of the condition bits upon which they depend. This variable is Boolean-valued. The default is false because the search space BC must explore is greatly expanded by this option. A reason you might set this variable to true might be that you were seeing excessive latency along a critical path that passed through the condition bit; or perhaps because the condition bit and the dependent operation(s) were being evaluated in the same cycle, and the total propagation delay along that path was excessive. You can force the dependent operations into previous cycles using the set_cycles family, if speculative execution is enabled.

The variable bc_fsm_coding_style is used to control the kind of control FSM that is generated. In early versions the control FSM was always built by FSM Compiler, but there was sometimes a problem where FSM Compiler would create false combinational cycles (which cause problems for gate-level optimization) if there were operations dependent on conditionals being performed in the same cycle as the conditionals. In later versions, BC supports the following set of options for this variable: one_hot builds a one-hot coded FSM without using FSM Compiler; this is usually the fastest and most expensive option. The counter_style option builds a minimum-length coded FSM without using FSM Compiler. The codes are assigned sequentially. This usually results in a less expensive but slower FSM. The two_hot option gives a compromise between one-hot and minimum-length codes. Finally, use_fsm_compiler (the default) uses the FSM compiler path that was the only option in earlier versions.

Notice that you can still set FSM Compiler options to select different coding styles; these will not result in the same circuits as the more recent options, because those options do not use FSM Compiler at all. Your choice of FSM style does not affect either schedule or allocation. However, in subsequent releases the BC team plans to provide capabilities that go beyond what is currently synthesizable, primarily in order to relax some of the constraints of fixed I/O mode; and when this happens FSM Compiler will be used only if the new features are not.

The variable **reset_clears_all_bc_registers**, when set to **true**, will connect the clear pins of all registers built by BC to the reset net. Otherwise, only registers that are read after a reset and before they are first written (e.g. the control FSM's state register and its saved-status bits) will be cleared by a reset pulse. Some users like to use this variable in conjunction with the **set_behavioral_async_reset** command so that all registers will be asynchronously cleared by a reset pulse.

The variables **vhdlout_sequential_delay** and **verilogout_sequential_delay** are used to get around a common problem in zero-delay simulation of the synthesized design. It often happens that such designs don't simulate properly, because the number of infinitesimal delays on a clock net turn out to be greater than the number of infinitesimal delays on the D or enable net of some register. The symptoms you will see in those circumstances are that the register appears to be being strobed one cycle too early. That isn't the case: it's just a race and the clock is arriving later than the data.

If you set the sequential delay variable to e.g. 2 ns before you write out the simulatable design, all Q pins will have the extra delay. That pushes the edges of registered signals back 2 ns, so the clock will have time to get to its destination before the data changes.

### Shell commands

The bc_shell command **set_margin** is used to control the margin allowed for control and muxing delays when timing the design before scheduling. Recall that the timing data generated before scheduling is used to estimate the delays and hence the achievability of chains of operations. But in the absence of the actual schedule and allocation, BC cannot know the delays attributable to multiplexing and control logic; if chained operations have short delay compared to that of a multiplexer, the 'unestimated' delays, and so the errors, can be large in proportion to the 'known' delays. BC addresses this by providing a fudge factor (the timing margin) which is added in to allow you to compensate for otherwise unestimated delays. The margin is expressed using the time units that are used in your technology library.

The command **register_control** (*q.v.*) is used to force registers on the inputs, outputs, or both of the control FSM. This can improve your cycle time at the expense of additional latency if the conditionals are on the critical path.

The command **set_stall_pin** is used to stop the design while waiting for some external event to occur. It functions by preventing the control FSM from advancing its state. The effect on your design is behaviorally indistinguishable from gating the clock, except that the reset will still work. This command is very useful when a pipelined loop needs to be synchronized with the outside world; recall that pipelined loops cannot contain other loops unless those other loops are unrolled.

## 5.4 Test Benches; Simulation

When you simulate the design, it is embedded in a *test bench* whose function is to stimulate and monitor the design during simulation. This section contains some tips on how to build good testbenches for use with BC.

### Clocking

In order to generate appropriate clocks, we have one or more clocking processes. These just generate square waves of the correct frequency.

### Reset

Resets pose a few special problems for simulation. BC registers all output signals of the synthesized design; this means that we incur a delay of one cycle between the nominal write (signal assignment) and when the new value actually becomes available on the signal (in Verilog, this is a port or a module-level **reg** variable). This means that for proper synchronization of the pre- and post-synthesis designs, you need to extend the reset pulse going to the pre-synthesis design by one cycle; so that in effect the pre-synthesis design 'begins to execute' one cycle after the post-synthesis design.

Another problem that comes up with resets is that you can specify a reset signal using only bc_shell commands. This means that your original HDL will not respond to the reset pulse; and important behaviors may only be simulated post-synthesis. You need to be aware of this and test the reset response of your circuit.

In most cases the reset pulse for the synthesized design only needs to be one cycle in duration. I have seen cases, however, in which the reset pulse needed to be as much as three or four cycles; this occurs because unknowns in the datapath registers propagate and don't get flushed out in a single cycle. This is the reason BC always clears the FSM's input status bits on reset: otherwise RTL simulation may never get out of an uninitialized state. At the gate level this is much less of a problem, but it still happens occasionally. The solution is just to extend the reset pulse by a cycle or two; that usually takes care of it. In a really severe case you might want to use **reset_clears_all_bc_registers**, but remember that this really only cures a simulation problem and it does cost gates.

### Data input

You will also want one or more data input processes. These can take data from a file or can generate it using more or less complicated methods. One simple way to get a 'heartbeat' test is to use a pseudorandom number generator; in DSP applications unit steps and impulses are often useful.

An important consideration in generating inputs is the timing under which the stimuli will be applied to the design under test. You want the inputs of the synthesized design to transition a little after the active clock edge, to avoid clock/data races. You also need to be careful about the times at which transitions occur if you have used superstate or free-floating mode: the reads may have moved, and so may see different inputs than in the original HDL. If your design incorporates handshaking, this is usually not a problem.

### Output monitors

You need one or more data output monitors. These watch the output signals and compare them to some 'golden' results, either computed, simulated (by a 'golden model'), or read in from a file. Again, the timing under which you look at the design under test's outputs is an important consideration. A simple output monitor can be nothing more than a VHDL assertion or a Verilog edge detector and conditional `$display`; but this is only the beginning.

There are often processes or circuits that represent other design modules. These may be RTL, gate-level, or behavioral; in general there will be a mix of the three and they will be in various stages of synthesis as your design progresses.

### Delays

It is always a good idea to have your stimulus data transitions occuring a short time after the rising clock edge. This avoids a problem with simultaneous data and clock transitions, in which both signals are retarded by some number of zero-time delays (also called *delta* delays), and whichever is retarded the least occurs 'first'. This kind of thing gets even more confusing in the behavioral context, because the original HDL is usually much less vulnerable to varying numbers of delta delays.

In the event that the output design is to be simulated at the RTL, or using only equational forms, or using zero-delay gate models, you should set the variable `vhdlout_sequential_delay` (or `verilogout_sequential_delay`) to a nanosecond or so. This will delay the output transitions of the synthesized design a little, and so avoid the problem in any network the synthesized design drives.

### Effect of I/O mode

The choice of I/O mode makes a big difference to the kind of test bench you write. In cycle-fixed mode, simulation of the pre- and post- synthesis designs should never differ at all; this leads to the simplest test bench. Here what you do is build a test bench that works for the input HDL; possibly modify the reset pulse's duration; insert the RTL-level design into the test bench in place of the input HDL; and then simulate. Alternatively, run the input and output designs simultaneously, with the same inputs, and compare the outputs on every active clock edge; meanwhile using the outputs of one or the other version to drive the rest of your circuit.

If you are using superstate or free-floating mode, interfacing becomes more difficult because the scheduler may have inserted extra clock edges, in effect slowing down the input design by adding more states. For this reason a test bench that depends on cycle counting to function correctly may not work in superstate or free-float mode. Instead of using cycle counts to synchronize communication between the test bench and the design, you must use handshaking; the handshaking can be more or less abstract depending on the specific situation. This is not as burden-some as it might at first appear; if the design has conditional state branches (e.g. as created by while loops) it will have some kind of explicit synchronization anyway. An example synchronization protocol is given in Chapter 8.

### Summary

This chapter has given you a set of tools that allow you to control the way BC schedules and allocates a design; in addition it describes a few settings that let you specify how the control FSM will be constructed. Also covered were some important aspects of designing test benches.

# Chapter 6

# Reports

This chapter describes the various reports that BC can generate during and after scheduling. We will concentrate on reports specifically generated by BC; other reports, e.g. of libraries used, cells used, area, and propagation delays, are described in the Design Compiler documentation.

There are three main classes of reports that BC can generate during and after scheduling. These are the timing and chaining report, generated after the design has been timed and before scheduling begins; the reservation table reports, which are generated at the user's discretion after scheduling; and the state machine reports, also generated at the user's discretion after scheduling. In addition, there are a number of more miscellaneous reports that deal with various error situations, and informational reports that tell you what is going on, issue warnings, and so on. We will not attempt to cover the reporting facilities of BC exhaustively, but will cover the most important reports and their uses in diagnosis and tuning. You may wish to refer to Chapter 7 as you read: it contains some full-length concrete examples.

## 6.1  Timing report

The first large report a BC user will see is the timing report generated after the design has been mapped and timed preparatory to scheduling. This report details the propagation delays of the components of the design. You will see it immediately after the timing analysis is done; recall that timing analysis is performed as part of either the `bc_time_design` command or the `schedule` command (see page 16).

Delays reported here pertain to chains of operations that might be constructed by scheduling. The chains result from data and control dependencies that are not broken by loop exits, forward skips, and so on. The chains are constructed so that each operation is at the beginning of a chain; the chains are as long as possible.

The delays are not reported on an operation-by-operation basis because a single lumped number is not accurate. A glaring example: consider two additions chained as shown on the left in Fig. 6.1. Now one could model the delays of the two additions separately; each might take, for example, ten nanoseconds. But looking at the right side of the figure, where two four-bit ripple-carry adders are shown in a chained configuration, it's easy to see that the chained operations do not in fact

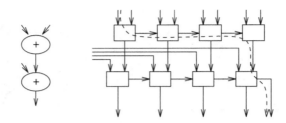

**Figure 6.1.** Two chained additions: delays don't add

have twice the delay of a single operation; the critical path (shown as a dashed line) is the carry chain of one of the adders with one additional full adder. Such circuits mean that BC must analyze timing starting at every data operation (because the wavefront of arrival times at the inputs of the operation may begin at a register output) and going through the adjacent operations (because the arrival times at those operations may differ depending on where the chain started).

Furthermore, because the clock period can be changed between timing analysis and scheduling, BC can't just cut off the analysis of a chain when the chained delay gets to be longer than the current clock cycle. Thus we must time chains beginning at all data operations and ending only at forced state transitions.

A fragment of a chaining report is shown on the right. It begins at an I/O read operation din_66. The read operation takes 0 ns because we assume that all inputs are externally registered with the same clock phase as the process currently under consideration. The chain then passes through a multiplication and three additions, and thence to an output operation dout_91.

```
Delay starting at din_66:
  din_66 = 0.000000
  mul_72 = 25.000401
    add_73 = 31.506996
      add_75 = 33.496998
        add_77 = 35.487000
          dout_91 = 35.48700
```

Another chain, taken from the same report, demonstrates graphically why we must analyze chains beginning at every operation, even if we have analyzed the operation before as part of another chain. In the chain shown above, the addition add_73 incrementally added about six nano-

```
Delay starting at add_73:
  add_73 = 19.596994
    add_75 = 21.586994
      add_77 = 23.576994
        dout_91 = 23.576994
```

seconds to the chain; but when the chain begins at **add_73**, its incremental delay is nearly twenty nanoseconds. Note, however, the small incremental delay associated with the further chaining of adders after **add_73**: each contributes only about two further nanoseconds.

This chain illustrates a case of data fanning out, i.e. chains that form a tree. The chain begins with **mul_69**, the result of which goes to both **add_77** and **add_70**. The fanout is conveyed by the indentation of the report.

```
Delay starting at mul_69:
  mul_69 = 27.610600
    add_77 = 40.340710
      dout_91 = 40.340710
    add_70 = 41.744106
```

The value of the chaining report is that you can set your clock period intelligently, if the clock period is a free parameter; alternatively, you can retarget your technology, if that is a free parameter; and finally, you can use the information to adjust the data dependencies and component implementations of the design. For example, you might see *multicycle operations* in your timing reports. Multicycle operations have a latency penalty because their inputs and control lines must be registered. The presence of multicycle operations is often, but not always, an indication that a better design can be achieved.

It is not always possible to *increase the clock period*; but if it is, you can usually find good trial clock periods by examining the chaining report. For example, in the report shown above, somewhere around 30 ns would be a good starting point, because of the multiplication `mul_69` which has a delay of 27.6 ns.

The chaining report can also help you to *set implementation constraints*. For example, you might want to put a `dont_use` attribute on the ripple carry adder; this would substantially change the numbers on the chaining report, and hence the scheduler's evaluation of the chains being considered. In the examples shown above, this would most likely be a bad idea, because of the deep chains of additions; but imagine a situation where an addition could be chained with a multiplication if the addition used a lookahead adder, but not with a ripple adder. You could then make chaining possible by setting `dont_use` on the ripple carry adder.

Recall from the discussion of section 2.2.1 that multicycle components are often inferior to pipelined components; the chaining report can tell you if multicycling is about to happen, and so that you should consider the use of a *pipelined component*.

*Adjusting data dependencies* is done by rewriting your source HDL. Generally speaking, you want to avoid deep chains, because they create long paths in the schedule, and so decrease BC's freedom to schedule operations in different cycles. But some chains are less important than others: as demonstrated by the chains of additions shown above. The chaining reports tell you where the data dependencies are, and what possible chains will be considered. These can often be manipulated by changing the order of arithmetic operations and operands, and by adding parentheses to your code. For example, the chains of additions shown above result from expressions of the form $a + b + c + d$. The parsing process builds such expressions as degenerate left to right associative trees, i.e. long chains. However, if we were to rewrite the same expression as $(a + b) + (c + d)$, the result would be a balanced tree. If the arrival times of one or more of the operands were substantially different from those of the other operands, we would rewrite the expression to give a shorter path length to the latest-arriving signal, and so on.

Finally, you can *set constraints* to break chains or force chains to be constructed. For example, the chaining analysis might tell you that two operations could not be chained within your clock cycle, but you might know that the chain could in fact be made to fit within the clock cycle, given a bit more effort in implementation selection, compilation, and mapping. In that case, you can force the chain to be constructed (using e.g. `set_cycles 0`) and let logic-level optimization take care of the apparent violation.

## 6.2   Reservation tables

The *reservation table* is a table whose columns represent abstract resources, and
whose rows represent csteps. This displays the schedule and allocation information
in a single unified form. It consists of rows, which correspond to csteps, and columns,
which correspond to resources. The reservation table is available after scheduling
and allocation have both been completed.

The main classes of resources listed in a reservation table are the registers and
the synthetic operations. In addition, BC lists I/O ports as resources; thus reads and
writes are displayed as well. Another special 'resource' that is displayed by default
is a column called 'Loops'; this column lists loop begin, end, exit, and continue
nodes. This column allows you to see the relationship between loop boundaries and
other operations. At the user's option, it is possible to display various subsets and
supersets of the resources and csteps that are displayed by default. For example,
one can display only a contiguous segment of all the csteps of the schedule; or
only resources of a certain type or name. These options allow you to decrease the
physical size of the report, and so help you to focus on the most interesting parts.

### 6.2.1   Operation Reports

```
bc_shell> report_schedule -operations
```

An example reservation table is shown on the opposite page. This example
has been edited to reduce its size; but otherwise it is as you would see it coming
from BC. The extreme left column shows the cycle (cstep) number. The next
column shows loop beginnings and ends, etc. The column after that is an input
port, followed by there are three arithmetic resources, then an output port. This
column ordering is the same in all operation reservation tables: cycles, loops, inputs,
operations, outputs. Displayed vertically above the operation columns are the types
of resources that have been allocated; here the resources are the ports, a Design-
Ware module DW01_add, and a pair of DW02_mults. Immediately below the resource
types are the resource names; here the resources are the port din, the adder r56,
and so on.

A separate table, an example of which you will find on page 109, gives lists
of resource types. One element of this table tells us that the adder r56 is a
(19_19->19)-bit DW01_add; this means it has two 19-bit inputs and a single 19-bit
output, and again gives its type.

In another table, the expanded names of the operations listed in the main reservation table are given. Hence main_design_loop_begin is the full name of the loop
operation L0. and a (16_16->16)-bit ADD_TC_OP operation is called o22 in the
table, while its full name is reset_loop/_L0/add_77. The numbers in parentheses
mean that the operation takes two 16-bit inputs and produce a 16-bit output, and
the TC_ADD means it is a twos-complement addition.

One thing that sometimes confuses users is the occurrence of multiple entries for
a single cstep and resource. This doesn't mean much when (as shown above) the

multiple entries are loop boundaries; just that the loop boundaries are somehow concurrent. But when the column having multiple entries in a single cstep is a physical resource, such as a multiplier, it means that the operations in that cstep and resource are conditional and mutually exclusive; thus only one of them can take place at a time.

```
                                        D       D
                                D       W       W
                                W       O       O
                                O       2       2
                                1       -       -
                        p       -       m       m       p
                        o       a       u       u       o
                        r       d       l       l       r
                        t       d       t       t       t
        -------+-----+-----+-----+-------+-----+------
        cycle | loop| din | r56 | r49   | r99 | dout
        -------------------------------------------------
           0   |..L3.|.....|.....|.......|.....|.W36..
               |..L0.|.....|.....|.......|.....|......
           1   |.....|.R41.|.....|.......|.....|......
           2   |.....|.R55.|.....|.......|.....|......
           3   |.L11.|.....|.o22.|..o26..|.o65.|......
               |..L6.|.....|.....|.......|.....|......
```

## 6.2.2  Variable Storage Reports

The command string shown here will generate a report that contains variable and CDFG data edge reservation information.

```
bc_shell> report_schedule -variables
```

The contents of the table are CDFG data arcs; again, these must be abbreviated in order to save space. An example of a complete variable reservation table can be found on page 112.

An important thing to know about variable reports is that data edges are reported over their lifetime, not over the time they are contained in storage. In the first cstep in which a data edge is displayed in a register's column, the edge is not in fact in the register; it is on the D pin of the register. Thus if a data edge $E$ originates at an operation occurring in cstep 6, then the first cstep in which it would appear in a variable report would be 6, and a read of the register during cstep 6 would see the previous contents of the register.

If the last use of the same variable was in cstep 10, then the lifetime of the variable would be from 6 to 10; the report of the register's contents would show the variable as present from 6 to 9, because some other variable might be on the D pin of the register in cstep 10.

The variable and operation names that are reported are those that BC uses to identify internal CDFG nodes and edges. They are unique in the design. Thus you can use these names for generating constraints, and they can be influenced by the use of labeling constructs as described in section 5.1. Variable names deserve a little more explanation, because they can be confusing.

Recall that the CDFG data edges represent particular data values, i.e. the contents of storage or the patterns of bits on a bus. They do not represent Verilog **reg** or wire variables or VHDL signals or variables; they represent the information contained in these constructs during various periods of time. Thus, for example, a variable initialized before a loop, set during the loop, and referenced after the loop would generate four different data edges: the initial value, the value contained in the variable during the first part of the loop, the value contained in the storage when the loop cycles back from its end to its beginning, and the value contained in the variable after the loop has exited. Each of these will be given a name constructed in a reasonable way; for example, the initial and loopback edges might be given names derived from the name of the variable, and the third edge might be given the name of the operation that created it, with the port identifier **Z** appended.

In addition, data edges created by elaboration will be stored in new storage resources in essentially arbitrary ways. Thus a data edge named after a variable will end up stored in a register with a completely different name; and another edge with a very similar provenance might be stored in the same register or in a different register, or even (if its producer and consumers are chained) not at all.

This all means that you should be careful when looking at variable reports: it is very easy to get confused about which data edge you are looking at, how the result of a particular operation is stored, and other similar questions. Similar confusions can arise when looking at operation names in the context of inlined or unrolled subprograms and loops; there we see many copies of the 'same' operation, each with a similar but not identical name.

## Masking

The **report_schedule** command supports a series of options designed to help you understand the synthesized design by reducing the sheer volume of text in the reports. These options allow you to reduce the numbers of rows and columns in any given report: thus allowing you to focus on particular aspects of the design.

The number of rows displayed can be controlled with the flags **-start** and/or **-finish**. These options take integer arguments and set the first and last csteps to be reported upon. The use of **-start** and **-finish** can also reduce the number of columns, because only columns with some content in the range of interest will be displayed. Hence, for example, if there is a resource **r40** that is unused in the interval between csteps 10 and 15, this command won't show **r40**'s column at all.

```
bc_shell> report_schedule -op -start 10 -finish 15
```

Columns can also be explicitly controlled. When reporting on variables, the additional flags `-min` and/or `-max` can be used. These set the range of register bit widths that will be displayed. Hence setting `-min 8` and `-max 12` will result in a display of only registers having bit widths between eight and twelve.

Another flag that you can use to control the columns displayed is the `-mask` flag. This flag takes as an argument a string of characters; each character permits a certain class of operations to be displayed. The alphabet of permissible mask characters when reporting operations is `rwLlop`. Of these, `r` and `w` cause I/O reads and writes to be displayed; `l` causes logic operations to be displayed, `o` causes synthetic operations to be displayed, `p` causes patch boxes to be displayed, and `L` causes loop boundaries to be displayed.

The default mask for `report_schedule -operations` is `rwoL`; this displays reads, writes, synthetic operations, and loops, as described above. Thus if you wanted only to see loop boundaries and I/O operations, you would use this.

```
bc_shell> report_schedule -op -mask rwL
```

### When to use reservation table reports

The reservation table is the best way to get a clear idea of the resources that BC created during allocation, and when and for what purposes they are used. It can be used to guide your exploration of the area/time tradeoff space. For example, you might look at the table and see that a resource was relatively underutilized, and that the operations allocated to it could be performed elsewhere with only slightly relaxed latency constraints. A reservation table will also help you to focus on the loops and data dependency paths that are critical in the sense of causing long latencies, as opposed to the critical paths in the sense of propagation delay, which are displayed in the timing report.

But the reservation table can be confusing when your purpose is to analyze what operations are competing with one another for the same resources, or if you want to find out the exact number of clock edges between one operation and another. Recall from our discussion of section 2.2.5 that csteps 'disappear' when loops, loop exits, or forward skips occur. Thus one might think that because operation $A$ was in cstep $i$ and operation $B$ was in cstep $j$, then the number of clock ticks between $A$ and $B$ would be $i - j$; but if there were loop boundaries or skips between the two, they could in fact be occurring in the same state.

Furthermore, a reservation table does not show any information about conditionals and conditionally executed operations. Thus you might easily see two additions being performed 'simultaneously' in one adder, as reported in the table; but they might be in completely different states, or be on different branches of a conditional. You would not be able to see this by looking at the reservation table. The state machine reports described in the next section are a way around this difficulty.

## 6.3   State Machine Reports

A tabular representation of the control FSM's state graph can be obtained by using either of two additional options of the **report_schedule** command. The flags that invoke these options are the **abstract_fsm** and **verbose_fsm** flags. The result of using these options is a state table: it shows states, conditions, next states, and outputs for the control FSM.

> bc_shell> report_schedule -abstract_fsm

The -**abstract** flag gives a concise description of the FSM's state table; the -**verbose** flag gives you more information. Here is an example report generated by the **abstract_fsm** option of **report_schedule**.

```
present           next
state   input   state       actions
------------------------------------------------------------
s_0_0   -       s_1_1       reset_loop/ready_35 (write)
                            reset_loop/dout_36 (write)
s_1_1   -       s_1_2       reset_loop/din_41 (read)
s_1_2   -       s_1_3       reset_loop/din_46 (read)
s_1_3   -       s_1_4       reset_loop/din_49 (read)
s_1_4   -       s_1_5       reset_loop/din_52 (read)
s_1_5   -       s_1_6       reset_loop/din_55 (read)
                            reset_loop/ready_56 (write)
s_1_6   -       s_3_7       reset_loop/_L0/_L1/start_60 (read)
                            reset_loop/_L0/mul_74 (operation)
                            reset_loop/_L0/mul_71 (operation)
s_3_7   1       s_2_9       reset_loop/_L0/din_66 (read)
                            reset_loop/_L0/ready_67 (write)
                            reset_loop/_L0/mul_72 (operation)
                            reset_loop/_L0/add_73 (operation)
s_3_7   0       s_3_7       reset_loop/_L0/_L1/start_60 (read)
```

The state table consists of four columns, which are the *state, input, next state,* and *actions* columns. These correspond exactly to the standard definition of a Mealy machine. So for example, if in state **s_3_7** the input is **1**, the next state will be **s_2_9** and the actions taken will be an I/O read, and I/O write, and two arithmetic operations.

If the input symbol on a line is a don't-care (-) then the transition (and associated operations) will occur no matter what's on the inputs of the FSM.

The **verbose_fsm** option is similar to **abstract_fsm** except in the amount of detail. For example, the action listed above as 'reset_loop/_L0/mul_72' would be listed as 'reset_loop/_L0/mul_72 in DW02_mult r99', and the other actions such as I/O reads and writes would also be expanded.

The abstract and verbose FSM modes of report_schedule use a mask flag similar to that of the variable and operation modes. The **r**, **w**, **l**, and **p** options have the same effect in the operation mode and in the FSM modes. In addition, the FSM modes support the following mask options that are not supported by the operation and variable options. The **d** character reports on data transfers in the design. This flag is only available when the verbose option is used. Using **s** reports the loop and cstep of each state. Using **t** reports on the CDFG status edges that drive the conditional branches. This is useful when doing desk analysis of flow of control in the FSM. Finally, **n** reports on the status nets being sampled; these correspond to the branches of the **t** flag, but can be shared in common status registers. This is more useful when monitoring a simulation, because you can monitor nets and you cannot monitor CDFG edges.

### When to use FSM style reports

If your design has a substantial amount of control branching, or if accounting for every clock tick is important, the FSM-style reports are more useful than the reservation table reports. This is because you can see where the operations are taking place on the state graph, and you can see how many transitions it takes to get from one state to another, and under what conditions a path will be taken.

The FSM-style report is also useful when you want to know exactly what the concurrency and conflict relationships between operations are. This shows in the assignment of operations to transitions; if two operations are on the same transition, or on two transitions whose sources and sinks are identical and whose conditions overlap, then those two operations are not mutually exclusive. You cannot infer this from the reservation table alone: two csteps may be different in the table, but they may be mapped to the same transition, and even when two operations take place in the same cstep, they may still have mutually exclusive branch conditions.

On the other hand, if you need a summary or overview of what is going on in your design, it can be difficult to extract from the FSM report, which does not show you edge lifetimes, resources, or resource usage directly.

## 6.4   Error messages

BC's error messages are largely self-explanatory. You can get manual information on them from BC itself, using the **man** command, or from the Synopsys documentation. If what's here differs from the documentation or what your version of BC actually does, trust the documentation. The purpose of this section is to give you a little supplementary background on a perplexing group of error messages: those that are emitted when BC cannot construct a schedule because of conflicting implicit and/or explicit constraints. These messages can be quite challenging.

In order to understand this class of error reports, you have to think about the order of events in scheduling. BC performs scheduling bottom-up; that is, from the innermost loop out. The loops are scheduled completely, i.e. there is no scheduling flexibility left in a loop when the scheduler finishes with it and pops up a level. Each time BC proceeds to the next loop, the inner loop is instantiated *as a single template* containing all of the contents of the inner loop. The template is rigid at that time and afterward: nothing can change it.

The loop $L$ that is currently being scheduled also contains templates. The template that represents the inner loop is just another of these. Thus the constraints on the contents of $L$ describe relationships between templates and operations.

When an inconsistency is detected, it may be such that some operation or template $T_1$ is supposed to fall after another template or operation $T_2$; and at the same time $T_1$ is also supposed to fall before or concurrently with $T_2$. One or both of the templates is derived from scheduling constraints on the structure of $L$; the other is often an inner loop's template, or an operation locked into that template.

When BC detects a contradiction of this kind it prints out the offending subgraph in as terse a way as it can, while still conveying the essential participants in the contradiction. This will often be expressed as a series of templates. You must then interpret these templates, *which do not exist in your source code,* to find out what things in your source code and manual constraints led to the contradiction.

Here's an example, shown in Verilog only because I will need to use line numbers.

```
 1   module con (clk, din, dout);
 2   input clk;
 3   input [7:0] din;
 4   output [7:0] dout;
 5   reg    [7:0] dout;
 6   always begin: main
 7      reg [7:0] tmp;
 8      forever begin
 9         @(posedge clk);
10         tmp = din;
11         @(posedge clk);
12         while (tmp > 8'h0) begin
13            dout <= din * (tmp + 8'h3);
14            @(posedge clk);
15            tmp = tmp - 8'h4;
16            while (tmp < 8'h20) begin
17               tmp = tmp + 8'h3;
18               @(posedge clk);
19            end
20            tmp = tmp + 8'h12;
21         end
22         @(posedge clk);
23      end
24   end
25   endmodule
```

The general strategy for dealing with the problem has three steps. First, find the correspondence between the things the report mentions into the source code.

This can't always be done easily; do as much as you can. Sketch the CDFG in the region of the problem. Again, do your best: the sketch may help, even if it is incomplete. Be sure you include the effect of any latency penalties you may be paying, e.g. because you have registered the control unit's inputs.

Second, draw the offending constraint graph and fill in the precedences between the source code objects, using the report as a guide.

Third, try to find a place where the contradiction can be resolved by removing or adding constraints or clock edge statements.

This source text is fed into **bc_shell** using the following **bc_shell** script:

```
analyze -f verilog contradiction.v
elaborate -s con
create_clock clk -p 10
schedule -effort zero
```

I used effort level zero because the question here is not quality but rather achievability. Zero effort lets us get to the core of the problem quickly; a higher effort level would serve no purpose.

The source as written cannot be scheduled in fixed I/O mode. The error message we will see is something like this, with minor variations in different versions of BC.

```
         Unsatisfiable Fixed Schedules found:
                (T_loop_16, 1) and
                (T_loop_12_design_loop_cont, 2)
            this violates min delay of 2 cycles
                 from T_loop_16
                 to T_loop_12_design_loop_cont
         Critical Paths from T_loop_16
                     to T_loop_12_design_loop_cont:
         Path 1
         loop_16                         (0)      0
          add_20                         (1)      1
            JOIN_L12_0                   (1)      2
              loop_12_design_loop_cont   (0)      2
         Path 2
         loop_16                         (0)      0
          add_20                         (1)      1
            JOIN_L12_0                   (1)      2
             JOIN_L12                    (0)      2
               loop_12_design_loop_cont (0)      2
         Path 3
         loop_16                         (0)      0
          add_20                         (1)      1
            loop_12_design_loop_cont     (1)      2
         Error: Fixed IO schedule is unsatisfiable (HLS-52)
```

What this report is telling us is that there is a fixed schedule (whether in fixed mode or not, the schedule is fixed by templates) and the fixed schedule is overconstrained, i.e. unsatisfiable.

The report is divided into two sections. The first section begins with "**Unsatisfiable fixed...**" and tells us that there is a fixed part of the schedule in which the templates or operations **T_loop_16** and **T_loop_12_design_loop_cont** are locked into steps 1 and 2, respectively. It also tells us that another constraint system wants them to be two cycles apart.

The second part of the report is a list of critical paths between the two named objects. These paths are the critical paths; if it were not for them the fixed schedule could be achieved.

So the first thing we have to do is identify the objects implicated in the constraint violation. The object **T_loop_16** is the template corresponding to the HDL loop beginning on line 16. That loop has already been scheduled; otherwise we would not be seeing details of the loop that contains it. Because it has been scheduled, its components are rigidly connected. Consulting the source HDL shows us that it must have length 2; there is only one clock edge in that loop.

The template **T_loop_16** is locked into cstep 1 of the current hierarchical context: we can tell that because of the number immediately following its name in the first part of the report.

The other object named in the report is **T_loop_12_design_loop_cont**. That is a template associated with the loop continue of the loop beginning on line 12. I'll call that loop 'L12' from here on. The loop continue template is locked into cstep 2 of L12; again, the number next to the template's name in the first part of the report tells us so.

The fact that the continue is named rather than the loop itself means that the conflict has occurred while scheduling L12.

The two templates are locked down by the numbers of clock edge statements between them and the loop beginning. **T_loop_16** is constrained to fall in cstep 1 of L12, i.e. its first cstep must line up with cstep 1 of L12. The continue's template is constrained to fall in cstep 2 of L12.

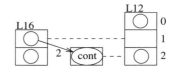

Referring to the source text, we see that there is one clock edge statement in L12, and it is between lines 12 and 16, which implies that the second loop must begin one cycle after the first. That confirms our impression that **T_loop_16** must be locked to cstep 1.

The next part of the first section of the message tells us that there is also a minimum delay of 2 cycles between the **T_loop_16** and the continue's template. Thus there is a contradiction and no schedule can be constructed.

The second part of the error message describes the critical paths that create the minimum delay of 2 cycles. There are three such paths. Each path is described as a sequence of nodes and/or templates, with the delay of the node/template in parentheses, and the accumulated path length in the right-hand column.

The first path goes from the beginning of the template **T_loop_16**, to the addition on line 20, to a join node that corresponds to the loop test and exit on line 12, and thence to the bottom (continue) node of the loop that begins on line 12. The addition is actually a red herring here: it is a proxy for the loop end of the loop beginning at line 16. The reason we don't see that loop end in the report is that it no longer exists: all that's left of the loop structure of that loop is the template, because *it has already been scheduled and inlined*. The add operation is just a convenient marker; it must occur one cycle after the beginning of **T_loop_16** (i.e. it can be chained with the end of the loop of line 16).

One cycle after the add is the earliest we can place the loop continue of the loop beginning on line 12.

We can't delete a cycle from the longer critical path; those constraints are there for a reason. What we *can* do is add a cycle into the outer loop in such a place that its template will be stretched a cycle; and so that there will be room for the 2-cycle constraint. That constraint is shown as a solid arrow on the opposite page.

So I'll insert a new clock edge in the Verilog, right after line 20. That will stretch the template of L12 in the right place.

```
 1   module con (clk, din, dout);
 2   input clk;
 3   input [7:0] din;
 4   output [7:0] dout;
 5   reg    [7:0] dout;
 6   always begin: main
 7      reg [7:0] tmp;
 8      forever begin
 9         @(posedge clk);
10         tmp = din;
11         @(posedge clk);
12         while (tmp > 8'h0) begin
13            dout <= din * (tmp + 8'h3);
14            @(posedge clk);
15            tmp = tmp - 8'h4;
16            while (tmp < 8'h20) begin
17               tmp = tmp + 8'h3;
18               @(posedge clk);
19            end
20            tmp = tmp + 8'h12;
21            @(posedge clk); // inserted here
22         end
23         @(posedge clk);
24      end
25   end
26   endmodule
```

After re-analysis and elaboration, I again create the clock and attempt to schedule it. It reports unsatisfiability again, for another reason. The new report looks like this.

```
Unsatisfiable Fixed Schedules found:
        (T_din_13, 0) and
        (T_dout_13, 0)
     this violates min delay of 3 cycles
          from T_din_13
          to T_dout_13
Critical Paths from T_din_13 to T_dout_13:
Path 1
din_13          (0)       0
 mul_13         (1)       1
   dout_13      (2)       3
```

Here the difficulty is located entirely on line 13, it seems. The relative schedule of the I/O writes din_13 and dout_13 is zero, because they fall between the same pair of clock edges; but the path length from the input to the output is three, because of the latency penalty (1) and the propagation delay (2) of the multicycle multiplication operation. Here's the timing report that says mul_13 is multicycle.

```
Cumulative delay starting at din_13:
    din_13  =  0.000000
    mul_13  =  20.126600
    dout_13  =  20.126600
```

The easiest solution to this class of problem is to relax the clock period. Setting it to 25 nanoseconds will get us to the next problem. Even if you can't freely change your clock period because of external design constraints, it might be a good idea to pretend that you can; it is often much easier to reduce the clock period of a design that can be scheduled than it is to try to add new clock edges to a design that can't be scheduled for other reasons.

But the multiplication's delay is not the last problem. The error message now reads like this.

```
Unsatisfiable Fixed Schedules found:
    (T_loop_12_design_loop_begin, 0)
    and (T_dout_13, 0)
  this violates min delay of 1 cycles
    from T_loop_12_design_loop_begin
    to T_dout_13
Critical Paths
    from T_loop_12_design_loop_begin
    to T_dout_13:
```

```
Path 1
loop_12_design_loop_begin         (0)        0
  add_13                          (0)        0
   mul_13                         (1)        1
    dout_13                       (0)        1
```

Here BC is evidently unhappy about the timing relationship of the loop beginning on line 12 and the output write on line 13. This is a problem for BC because at least one cycle must come between the loop beginning (and the loop test), and any I/O operation that depends on the loop test. The clue here is that BC is saying that the fixed-I/O schedule calls for zero cycles, meaning that the number of clock edge statements between the two constructs is zero. The contradiction is that there is a minimum delay of one cycle between the loop begin node and the write node; this happens because there is a loop test and a conditional state branch to be resolved before the write (or any I/O) can be committed[1].

So I'll transpose the clock edge on line 14 with the output statement on line 13.

```
1    module con (clk, din, dout);
2    input clk;
3    input [7:0] din;
4    output [7:0] dout;
5    reg    [7:0] dout;
6    always begin: main
7       reg [7:0] tmp;
8       forever begin
9          @(posedge clk);
10         tmp = din;
11         @(posedge clk);
12         while (tmp > 8'h0) begin
13            @(posedge clk);
14            dout <= din * (tmp + 8'h3);
15            tmp = tmp - 8'h4;
16            while (tmp < 8'h20) begin
17                tmp = tmp + 8'h3;
18                @(posedge clk);
19            end
20            tmp = tmp + 8'h12;
21            @(posedge clk);
22         end
23         @(posedge clk);
24      end
25   end
26   endmodule
```

---

[1] A conditional that did not imply a state branch would not be affected by this consideration.

Finally the schedule is satisfiable in fixed I/O mode. Now I can go back and start optimizing the design. For example, I really should go back to a ten-nanosecond clock; if I do, I will have to separate the read and write of line 14 with clock edge statements so that there is time to do the multiplication between the read and the write. To do this, I will have the following choices. First, I can break the multiplication down into several smaller multiplications and some additions. Thus each constituent operation will fit into the shorter clock cycle. Second, I can allow the multiplication to be multicycled (section 2.2.1). Third, I can try to use logic level optimization to force the multiplication to fit into a single cycle; see section 5.3. Fourth, I can pipeline the multiplication; section 2.2.1 describes that option. But unless I can somehow force the multiplication into a single cycle, I will need to put in some clock edges between the read of din and the write of dout; it is just a question of how many, and what the hardware will cost.

Another thing I could do now would be to somehow reorder my code to need less latency. This is highly problem-dependent; you have to think about what you are trying to accomplish. For example, a pipelined loop, if you can use it, often seems miraculous in the way it raises throughput while sometimes even lowering cost; but sometimes it doesn't do you much good because of the structure of the CDFG.

Unrolling and rolling loops in different ways also helps, sometimes; as does reordering arithmetic operations. The trick is to think of different ways to write the code that will achieve the same functional effect but that BC will be able to interpret differently.

### Summary

In this chapter we have seen a brief introduction to the reports that BC generates. The discussion took examples from the current version of BC; your reports will vary, perhaps in significant ways. But one thing will remain constant: the need to think about how BC perceives your design, and the constraints that it must obey.

The next chapters give examples of designs and reports, and illustrate the thought process that leads us from the reports to improvements in the designs.

# Chapter 7

# FIR filter

This chapter describes the HDL text for and synthesis of a finite impulse response (FIR) filter of 17 taps. We will use superstate-fixed mode scheduling and manual constraints to explore the space of achievable designs.

The main purpose of this chapter is to present a coherent and complete example. The secondary purpose is to demonstrate superstate-mode scheduling. The first synthesis run results in a fast, large design; succeeding runs show how to decrease the cost at the expense of performance, thus exploring part of the achievable space for this particular filter.

FIR filters are common in modern design practice because they are stable, efficient, and easy to design. The filter operates on a stream of input samples $u[k]$, where $k$ is an indexing variable that describes the position of a particular sample in the input stream. It produces a stream of results $y[k]$:

$$y[k] = \sum_{i=0}^{L} h[i]u[k-i]$$

The filter is defined for a quasi-infinite stream of samples indexed by the variable $k$ (by convention, $k \geq 0$). For some fixed $L$, which is the number of *taps* of the filter, here 17, it looks at the last $L$ samples and multiplies them by coefficients $h[i]$, which are constants that determine the filter's characteristic function.

## 7.1  Initial Design

The first thing we need for our filter is the coefficient array. In VHDL this can be defined as a constant type in a package; the Verilog equivalent will have to wait until we define a **module**. This set of coefficients, which was generated by COSSAP, defines a reasonable low-pass filter [1].

---

[1] Notice that these coefficients are *symmetric*: i.e. the last coefficient is the same as the first, and so on. This would normally lead us to an optimized algorithm of half as many multiplications.

```
library ieee;
use ieee.std_logic_1164.all;
use ieee.std_logic_arith.all;
package coeffs is
  type coef_arr is array (0 to 16) of signed (8 downto 0);
  constant coefs: coef_arr := coef_arr'(
      "111111001", "111111011", "000001101",
      "000010000", "111101101", "111010110",
      "000010111", "010011010", "011011110",
      "010011010", "000010111", "111010110",
      "111101101", "000010000", "000001101",
      "111111011", "111111001");
  end coeffs;
```

Now we will define the pinout of the filter as a VHDL **entity**. Again, we defer the Verilog until we can speak of the **module**.

```
library ieee;
use ieee.std_logic_1164.all;
use ieee.std_logic_arith.all;
use work.coeffs.all;

entity fir is
 port (clk, reset: in std_logic;
       sample: in signed (7 downto 0);
       result: out signed (9 downto 0));
end fir;
```

The filter has a clock input and a reset input: each is one bit wide. I normally use **std_logic** when defining VHDL signals and variables; other types are definable and usable, but this is a nice arithmetic package (the **library** and and **use** statements above) that behaves well in synthesis and simulation. Note also that I had to **use** the package of coefficients.

The architecture of the filter is called 'beh'. It has a single process. There are a few local variables of various bitwidths; of these, the variable **shift** is the most interesting, because it is an array of 8-bit words. This array holds the latest $L + 1$ inputs $u[k - i]$; in each sample cycle the array is shifted by one word.

Notice that I didn't map **shift** to a memory. That would have severely restricted the performance of the filter, because the access bandwidth of the memory would have become a bottleneck.

But because I did not map it to a memory, **shift** will be elaborated as a single large (136-bit) variable. If such a variable is used incautiously, a number of very wide data edges could be created in the CDFG. If the data edges don't have disjoint lifetimes, they need separate storage; this in turn may result in a very large register cost. So it is important in cases like this to be careful about the variable lifetimes.

Another thing that can happen if I am unlucky or incautious is a large multiplexing cost. This occurs when a new CDFG edge is written into the register, and the new edge's definition is the same as the old one's except for a field whose index is determined at run time. That is, expensive variable writes are generated by expressions of the form

```
shift[index] = data;
```

where **index** is not a compile-time constant.

Alternatively, I could have defined the shift register as a collection of individual variables. That would have had a bad impact on the elegance of the specification and/or my ability to use a loop, but it would have prevented the creation of 136-bit CDFG edges and still preserved a high access bandwidth.

Nevertheless, by being careful we can define the big edges in such a way that they are read and written only once per sample and have disjoint lifetimes; then the register costs should be acceptable.

```
architecture beh of fir is
begin
  fir_main: process
      type shift_arr is array (16 downto 0)
                         of signed (7 downto 0);
      variable tmp, old: signed (7 downto 0);
      variable pro: signed (16 downto 0);
      variable acc: signed (18 downto 0);
      variable shift: shift_arr;
  begin
    reset_loop: loop
        for i in 0 to 15 loop -- zero out the shift register
            shift(i) := (others => '0');
        end loop;
        result <= (others => '0');
        wait until clk'event and clk = '1';
        if reset = '1' then exit reset_loop; end if;
        main: loop
            tmp := sample;
            pro := tmp * coefs(0);
            acc := conv_signed(pro, 19);
            for i in 15 downto 0 loop
                old := shift(i);
                pro := old * coefs(i + 1);
                acc := acc + conv_signed(pro, 19);
                shift(i + 1) := shift(i);
            end loop;
            shift(0) := tmp;
            result <= acc(18 downto 9);
```

```
                wait until clk'event and clk = '1';
                if reset = '1' then exit reset_loop; end if;
            end loop main;
        end loop reset_loop;
    end process;
end beh;
```

In Verilog, I will begin by defining a macro `'wait` which performs the clock edge functions. The simulator may not allow the line continuation characters; you may have to delete the line feeds that make this macro readable.

```
'define wait                                             \
    begin                                                \
      @(posedge clk);                                    \
      if (reset == 1'b1) disable fir_main;               \
    end
```

The Verilog for the filter is a module with four ports: two one-bit inputs **clk** and **reset**, an eight-bit input **sample**, and a ten-bit output **result**.

The large shift register can be defined as either a 136-bit **reg** variable in the **always** block, or as a Verilog memory in the scope of the module. I declared it as a global in order to keep the rest of the code clean; but it makes no real difference in the circuit.

Notice that merely using the Verilog memory syntax does not imply a memory in BC. In order to do that we must explicitly bind the variable to a DesignWare memory. Note also that I had to declare the Verilog memory in the module's scope rather than that of the **always** block: if I wanted to use an actual RAM in the implementation, I would have had to put the declaration in the **always** block.

The coefficient array can be represented for synthesis by a parameter representing a big array of bits. Unfortunately, the Verilog simulator won't accept nonconstant expressions that index into an array of bits; so I had to redefine it as a Verilog memory and load it explicitly. That was too ugly to print (But it is included it on the diskette, named **ugly.v**). In addition, the result of synthesis would be a register instead of a bundle of constants; you'd have to fix that by hand at a lower level.

Another way I could have approached the coefficient array is to have used accesses a single bit wide. Such array accesses can use nonconstant index expressions. I would have had to define a new loop (which I would unroll; there would be little point to serializing the parameter access).

Another feature of the Verilog language is its lack of signed arithmetic facilities; you must define a signed multiplication operation yourself. Do this using a function and then map the function to the signed DesignWare multiplier using the **map_to_operator** pragma shown here.

```
module fir (clk, reset, sample, result);
input clk, reset;
input  [7:0] sample;
output [9:0] result;
reg    [9:0] result;
/* a bug in BC v3.4a forces this out of the always block */
reg    [7:0] shift [0:16];
    function [16:0] mul_tc;
    /* this is needed for twos complement multiplication */
    input [7:0] A; // the port names A, B, Z
    input [8:0] B; // are taken from the designWare
                   // otherwise this won't work
    reg sgn;
      // synopsys map_to_operator MULT_TC_OP
      // synopsys return_port_name Z
      // the above tells how to map this whole function
      // to one designWare function.
    begin
      // the following is only for simulation
      sgn = A[7] ^ B[8];
      if (A[7] == 1'b1) A = ~A + 1'b1;
      if (B[8] == 1'b1) B = ~B + 1'b1;
      mul_tc = A * B;
      if (sgn == 1'b1) mul_tc = ~mul_tc + 1'b1;
    end
  endfunction

  always begin: fir_main
    reg [5:0] i;
    reg [7:0] tmp, old;
    reg [16:0] pro;
    reg [18:0] acc;
    /* this is the pretty version */
    parameter coefs =
        { 9'b111111001, 9'b111111011, 9'b000001101,
          9'b000010000, 9'b111101101, 9'b111010110,
          9'b000010111, 9'b010011010, 9'b011011110,
          9'b010011010, 9'b000010111, 9'b111010110,
          9'b111101101, 9'b000010000, 9'b000001101,
          9'b111111011, 9'b111111001 };
    forever begin: reset_loop
      for (i = 6'h0; i < 6'd17; i = i + 6'h1) begin
        shift[i] = 8'h0;
      end
```

```
            result <= 10'h0;
            'wait
            forever begin: main
               acc = 18'h0;
               tmp = sample;
               pro = mul_tc(tmp, coefs[8:0]); // tmp * coefs[0]
               acc = { pro[16], pro[16], pro };
               for (i = 6'd16; i > 6'd0; i = i - 6'h1) begin
                   old = shift[i - 1];
                   pro = mul_tc(old, coefs[i * 9 + 8 : i * 9]);
                           // old * coefs[i]... this synthesizes but
                           // check the Verilog standard on [x:y],
                           // it's outside the language!
                   acc = acc + { pro[16], pro[16], pro };
                   shift[i] = old;
               end
               shift[0] = tmp;
               result <= acc[18:9]; /* synopsys line_label aut */
               'wait
            end
         end
      end
endmodule
```

The design itself begins with the reset tail, in which the shift register is zeroed out. This might not be necessary in some designs: because the filter has a limited response, any transients that occur during power-up will eventually die out. Notice that the first **for** loop will be unrolled by default; that's fine, all we really want is for the shift register to be zeroed out.

After that we have the beginning of the main loop. The first thing that happens in the main loop is that the accumulator **acc** is zeroed; this is where we will store the summed result. The sample input is read and the first element of the sum is computed. Saving this value until after all the other products had been computed and summed, and only then adding it into the sum, might be a good idea: think about the shape of and constraints on the CDFG.

The innermost loop will also be unrolled; it isn't protected by a pragma. In the Verilog as given above, leaving the loop rolled would cause a big increase in area, due to the addressing logic needed; you would be much better off declaring the coefficient array as a Verilog memory and explicitly loading its contents if you intended to leave it rolled. However, this has another problem: the memory will be mapped to a big register, roughly doubling your register cost. So you would then have to create a ROM binding for the coefficient array. In addition, the use of a single big parameter is problematic in that Verilog simulators don't support slices with indices whose values are other than constant expressions (Verilog synthesis doesn't really support it either. But in an unrolled loop the loop index 'variable' becomes a series of constants.).

In the innermost loop the first fifteen elements of the shift register are multiplied by the appropriate coefficients and then shifted. The last sample falls off the end and is never used. I did it this way because the shifter loop is made a bit simpler by letting the last sample drop off the end. Don't forget that logic compilation will delete storage whose output is never used; if you look closely at the innermost loop, you'll see that in fact the top word of the shifter is going to go away for this reason.

The result is then written out from the top bits of the accumulator; and a clock edge separates the write from the bottom of the loop. Otherwise this design would be illegal because of the read operation in the first superstate of the loop.

Notice that the Verilog includes a line label on the line that writes out the result. The line label is `aut`; that name will be given to all of the constructs on the line (here, only the write operation).

## 7.2   Synthesis

The filter is now ready for scheduling. Here is a bc_shell script. I included commands to parse both the Verilog and the VHDL versions; but the script can only do one. I commented out the Verilog, to avoid the confusion that the line continuation characters in the `'wait` macro can cause: the routine that assigns operation names in BC doesn't see them. In any case the Verilog and VHDL descriptions lead to substantially similar results.

The technology library was `lsi_10k`; this is a convenient and stable library, good for examples. You might want to try a different library; but expect different results if you do.

```
analyze -f vhdl fir.vhd          /* parse the VHDL */
/* analyze -f verilog fir.v */   /* parse the Verilog */
elaborate -s fir                 /* elaborate */
create_clock clk -p 50           /* create the clock */
schedule -io super
report_schedule -summary        > sum_rep
report_schedule -op -mask rwL   > rwl_rep
report_schedule -op -mask rwo   > rwo_rep
report_schedule -abs            > fsm_rep
report_schedule -var            > var_rep
/* Now write out the RTL */
/* write -hier -f verilog -out fir_scheduled.v */
write -hier -f vhdl -out fir_scheduled.vhd
/* uniquify and compile from here */
```

The circuit that is generated is summarized in the following report. I have edited it only to the extent of shortening some lines to fit it into this page format; I will edit subsequent examples more, but I wanted to give one full-length example.

```
bc_shell> report_schedule -sum

**********************************************************************
                Date      : Tue Feb 20 07:36:03 1996
                Version   : v3.4a
                Design    : fir
**********************************************************************


*******************************************
*  Summary report for process fir_main:  *
*******************************************
--------------------------------------------------------------------
        Timing Summary
--------------------------------------------------------------------
 Clock period 50.00
 Loop timing information:
     fir_main...........................3 cycles (cycles 0 - 3)
           reset_loop.....................3 cycles (cycles 0 - 3)
                 main......................2 cycles (cycles 1 - 3)
--------------------------------------------------------------------
        Area Summary
--------------------------------------------------------------------
 Estimated combinational area    7029
 Estimated sequential area       1920
 TOTAL                           8949

 4 control states
 4 basic transitions
 2 control inputs
 5 control outputs


--------------------------------------------------------------------
        Resource types
--------------------------------------------------------------------
        Register Types
========================================
     8-bit register....................1
     17-bit register...................1
     19-bit register...................1
     136-bit register..................1

        Operator Types
========================================
     (4_8->12)-bit DW02_mult...........1
```

```
    (8_4->12)-bit DW02_mult............1
    (8_5->13)-bit DW02_mult............1
    (8_8->12)-bit DW02_mult............1
    (8_8->14)-bit DW02_mult............2
    (8_8->15)-bit DW02_mult............1
    (9_8->17)-bit DW02_mult............2
    (19_19->19)-bit DW01_add...........8

    I/O Ports
========================================
    8-bit input port...................1
    10-bit registered output port......1


----------------------------------------------------------------
```

The report is divided into three main parts: the timing summary, the area summary, and a resource type summary. The timing summary tells us that the main loop has a duration of three cycles (recall that one control step will disappear from a rolled loop); that the reset loop takes three cycles; and that the inner loop has a duration of two cycles. For the clearest idea of the loop latencies you should consult the FSM-style report (shown below); the summary report is indicative but not completely authoritative, because of csteps that can disappear.

The area report gives us estimated combinational, sequential, and control unit figures. The combinational area includes functional units such as multipliers, multiplexers, and whatever random logic is included in the datapath. The control unit area estimates are more abstract: there are four states and four transitions between the states, and the FSM has two inputs and six outputs. The term 'basic' as applied to transitions means just that there are four unique combinations of states and next states; this is a rough indicator of control FSM complexity.

The resource summary tells us that we have three registers, totaling 180 bits of storage; this is exclusive of the FSM's state bits and the output register. The synthetic operators used are eight multipliers and eight adders.

In addition to the registers and the synthetic components, there are two ports in the report. An input port costs no gates, at least until layout is done, when it may or may not involve extra silicon; the output port is registered, which contributes to sequential area but is not reported as a register.

The next report gives a clearer idea of the timing. I have used the **-mask rwL** option, which shows only loop boundaries, reads, and writes. This is a good way to get a rough idea of the black box behavior of the synthesized design. If we had used the default mask, there would be so many synthetic (arithmetic) operations in this design that the report would not easily fit on the page.

It isn't all that good for finding out exactly what the interfacing timing is, though: use the FSM-style reports for that.

```
bc_shell> report_schedule -op -mask rwL
**********************************************
*  Operation schedule of process fir_main:  *
**********************************************

    Resource types
=======================================
    loop......loop boundaries
    p0........8-bit input port sample
    p1........10-bit registered output port result


                    p     p
                    o     o
                    r     r
                    t     t
    -------+------+-----+-----
    cycle | loop | p0  | p1
    -------------------------
      0   |..L3..|.....|.W40.
          |..L0..|.....|.....
      1   |..L6..|.R44.|.....
      2   |......|.....|.W54.
      3   |..L8..|.....|.....
          |..L7..|.....|.....
          |..L5..|.....|.....
          |..L4..|.....|.....
          |..L2..|.....|.....
          |..L1..|.....|.....


    Operation name abbreviations
=============================
    L0.......fir_main_design_loop_begin
    L1.......fir_main_design_loop_end
    L2.......fir_main_design_loop_cont
    L3.......reset_loop/reset_loop_design_loop_begin
    L4.......reset_loop/reset_loop_design_loop_end
    L5.......reset_loop/reset_loop_design_loop_cont
    L6.......reset_loop/main/main_design_loop_begin
    L7.......reset_loop/main/main_design_loop_end
    L8.......reset_loop/main/main_design_loop_cont
    R44......8-bit read reset_loop/main/sample_44
    W40......10-bit write reset_loop/result_40
    W54......10-bit write reset_loop/main/result_54
```

This report shows the loop boundaries L0 – L8 and the I/O operations R44, W40, and W54. Note that the data read R44 and the result write W54 are in adjacent csteps; thus the entire computation is taking place in those two cycles. W40 can be ignored; it is just initializing the output.

The next table shows timing and resource usage; notice that I had to fold the table and delete two columns (an adder and a multiplier) to make it fit. This is the operation reservation table. I used the **-mask rwo** option; this shows only the reads, writes, and synthetic operations. I had to split the table into three groups in order to make it fit; and I have edited it a bit to make it more concise as well.

```
report_schedule -op -mask rwo

    Resource types
=====================================
    p0.........8-bit input port sample
    p1.........10-bit registered output port result
    r53........(19_19->19)-bit DW01_add
    r55........(19_19->19)-bit DW01_add
    r56........(19_19->19)-bit DW01_add
    r57........(19_19->19)-bit DW01_add
    r58........(19_19->19)-bit DW01_add
    r59........(19_19->19)-bit DW01_add
    r66........(8_8->15)-bit DW02_mult
    r1166......(9_8->17)-bit DW02_mult
    r1180......(8_8->14)-bit DW02_mult
    r1243......(9_8->17)-bit DW02_mult
    r1277......(8_8->14)-bit DW02_mult
    r1321......(8_8->12)-bit DW02_mult
    r1343......(8_5->13)-bit DW02_mult
    r2115......(19_19->19)-bit DW01_add
    r2116......(19_19->19)-bit DW01_add
    r2466......(4_8->12)-bit DW02_mult
    r2469......(8_4->12)-bit DW02_mult
```

| | port | add | add | add | add | add | add |
|---|---|---|---|---|---|---|---|
| cycle | p0 | r58 | r55 | r2116 | r56 | r2115 | r59 |
| 0 | ..... | ....... | ....... | ....... | ....... | ....... | ....... |
| 1 | .R44. | .o686.. | .o686h. | .o686c. | .o686d. | .o686g. | .o686n. |
| 2 | ..... | .o686b. | .o686a. | .o686k. | .o686l. | .o686e. | .o686f. |
| 3 | ..... | ....... | ....... | ....... | ....... | ....... | ....... |

```
             add       add      mult      mult      mult      mult      mult
      -------+-------+-------+-------+-------+-------+-------+-------
      cycle |  r57  |  r53  | r1166 | r1243 | r1321 | r2469 | r2466
      -------+-------------------------------------------------------
        0    |.......|.......|.......|.......|.......|.......|.......
        1    |.o686o.|.o686j.|.o605..|.o605b.|.o605d.|.o605e.|.o605f.
        2    |.o686i.|.o686m.|.o605a.|.o605l.|.o605c.|.......|.......
        3    |.......|.......|.......|.......|.......|.......|.......
```

```
             mult      mult      mult      mult     port
      -------+-------+-------+-------+-------+-----
      cycle | r1343 | r1180 |  r66  | r1277 | p1
      ---------------------------------------------
        0    |.......|.......|.......|.......|.W40.
        1    |.o605g.|.o605h.|.o605i.|.o605j.|.....
        2    |.......|.o605m.|.o605k.|.o605n.|.W54.
        3    |.......|.......|.......|.......|.....
```

Operation name abbreviations
==========================
```
R44....8-bit read reset_loop/main/sample_44
W40....10-bit write reset_loop/result_40
W54....10-bit write reset_loop/main/result_54
o605...(8_9->17)-bit MULT_TC_OP reset_loop/main/mul_49_i7
o686...(19_19->19)-bit ADD_TC_OP reset_loop/main/add_50_i13
o605a..(8_9->17)-bit MULT_TC_OP reset_loop/main/mul_49_i6
o605b..(8_9->17)-bit MULT_TC_OP reset_loop/main/mul_49_i8
o605c..(8_4->12)-bit MULT_TC_OP reset_loop/main/mul_49_i0
o605d..(8_4->12)-bit MULT_TC_OP reset_loop/main/mul_45
o605e..(8_4->12)-bit MULT_TC_OP reset_loop/main/mul_49_i15
o605f..(8_4->12)-bit MULT_TC_OP reset_loop/main/mul_49_i14
o605g..(8_5->13)-bit MULT_TC_OP reset_loop/main/mul_49_i13
o605h..(8_6->14)-bit MULT_TC_OP reset_loop/main/mul_49_i11
o605i..(8_7->15)-bit MULT_TC_OP reset_loop/main/mul_49_i10
o605j..(8_6->14)-bit MULT_TC_OP reset_loop/main/mul_49_i9
o605k..(8_6->14)-bit MULT_TC_OP reset_loop/main/mul_49_i5
o605l..(8_7->15)-bit MULT_TC_OP reset_loop/main/mul_49_i4
o605m..(8_6->14)-bit MULT_TC_OP reset_loop/main/mul_49_i3
o605n..(8_5->13)-bit MULT_TC_OP reset_loop/main/mul_49_i1
o686a..(19_19->19)-bit ADD_TC_OP reset_loop/main/add_50_i2
o686b..(19_19->19)-bit ADD_TC_OP reset_loop/main/add_50_i5
o686c..(19_19->19)-bit ADD_TC_OP reset_loop/main/add_50_i14
o686d..(19_19->19)-bit ADD_TC_OP reset_loop/main/add_50_i15
```

```
o686e..(19_19->19)-bit ADD_TC_OP reset_loop/main/add_50_i4
o686f..(19_19->19)-bit ADD_TC_OP reset_loop/main/add_50_i3
o686g..(19_19->19)-bit ADD_TC_OP reset_loop/main/add_50_i12
o686h..(19_19->19)-bit ADD_TC_OP reset_loop/main/add_50_i10
o686i..(19_19->19)-bit ADD_TC_OP reset_loop/main/add_50_i1
o686j..(19_19->19)-bit ADD_TC_OP reset_loop/main/add_50_i8
o686k..(19_19->19)-bit ADD_TC_OP reset_loop/main/add_50_i6
o686l..(19_19->19)-bit ADD_TC_OP reset_loop/main/add_50_i7
o686m..(19_19->19)-bit ADD_TC_OP reset_loop/main/add_50_i0
o686n..(19_19->19)-bit ADD_TC_OP reset_loop/main/add_50_i11
o686o..(19_19->19)-bit ADD_TC_OP reset_loop/main/add_50_i9
```

This table shows us the overall utilization of the functional units and the I/O behavior. The functional units are only fairly well utilized; three of the multipliers are used in only one cycle of the two available. This suggests that we could eliminate one of them by moving an operation a little later. When I tried it with a longer clock period I was able to force that solution using **set_cycles**; but I couldn't achieve it with the short clock cycle. The conclusion I drew was that chaining depth is the problem here. That is, it is necessary to perform most of the multiplications early, so that the summing tree can be finished up in the next clock cycle.

Notice also that there are only fifteen multiplications listed here. The filter has seventeen taps; one might wonder where the other two multiplications went. The answer is evident when we take a closer look at the coefficients: two of them are powers of two. That means they can be done using left shifts; and in fact we don't even need a shifter because they can be hardwired. BC performs this optimization automatically.

The next report we'll look at is the variable usage report. Recall from Chapter 6 that the variables are displayed from their date of birth to the cstep in which they are overwritten.

```
bc_shell> report_schedule -var

*****************************************
*  Register usage of process fir_main:  *
*****************************************

    Storage resource types
========================
    r3.........136-bit register
    r52........8-bit register
    r1330......19-bit register
    r1339......17-bit register
```

```
-------+-------+-------+-------+-----
cycle |  r3   | r1330 | r1339 | r52
---------------------------------------
      | (136) |  (19) |  (17) |  (8)
=======================================
  0   |.......|.......|.......|.....
  1   |..v0...|..v1...|..v2...|.v3..
  2   |..v0...|.......|.......|.v3..
  3   |.......|.......|.......|.....
```

```
Data value name abbreviations
=====================
v0......136-bit data value reset_loop/main/shift
v1......19-bit data value reset_loop/main/add_50_i8/Z
v2......17-bit data value reset_loop/main/mul_49_i7/Z
v3......8-bit data value reset_loop/main/sample_44/net
```

The top of this report shows the register population (which does not include output port registers). The center shows the CDFG edges and their lifetimes; for example, the CDFG edge **v0** is born in cstep 1 and dies in cstep 2. The long names for the edges are given in the lower table; sometimes the bit widths of the edges are less than the widths of the registers they are stored in, but that is not the case here. The values being stored are, in order, the contents of the shift register; two intermediate values in the summation; and the input sample. Given that the whole could not be computed in one cycle, this is about as good as we can reasonably expect. The conclusion that we can draw from looking at this report is therefore that it isn't worth trying to reduce the register count.

Finally, let's look at the FSM state table. Again, I have edited the report to make it fit; but the essentials are all here.

This report I generated using the **-abstract** flag and the mask option. The mask I used was **rw**. The result of this mask option is that only reads and writes will be displayed; this cuts down the size of the report substantially and lets us see the process's interface as a state table. This mask option is the best for deriving timing diagrams and external (black box) behavior of the design.

```
bc_shell> report_schedule -abs -mask rw
```

```
present          next
state  input     state       actions
---------------------------------------------------------------
s_0_0    -        s_1_1     a_2: reset_loop/result_40 (write)
s_1_1    -        s_2_2     a_0: reset_loop/main/sample_44 (read)
s_2_2    -        s_2_3     a_1: reset_loop/main/result_54 (write)
s_2_3    -        s_2_2     a_0: reset_loop/main/sample_44 (read)
---------------------------------------------------------------
```

The finite-state machine style report gives the clearest idea of what the timing of the design really is. The report shows four states: s_0_0, s_1_1, s_2_2, and s_2_3. The first two states are the reset tail. This is apparent because there is no path to the first state, s_0_0; hence it must be reachable only by asserting a reset. The second two states are the main loop. Note, however, that the read occurs on both the transition s_1_1 – s_2_2 and s_2_3 – s_2_2; the write occurs on the transition s_2_2 – s_2_3. There is no branching in this filter, so the conditions of the state transitions are always '-' (don't care).

Here is a report that includes the synthetic operations (adds and multiplies) as well as the reads and writes. As you can see, it is a little more difficult to interpret the state graph when you don't use the mask; but then, the previous example didn't show what was going on inside the design (imagine writing this all out by hand, using RTL techniques!).

```
bc_shell> report_schedule -abs


**************************************************
*   State graph style report for process fir_main:  *
**************************************************
==================================================================
present         next
state   input state     actions
------------------------------------------------------------------
s_0_0    -      s_1_1     reset_loop/result_40 (write)
s_1_1    -      s_2_2     reset_loop/main/sample_44 (read)
                          reset_loop/main/add_50_i13
                          reset_loop/main/add_50_i14
                          reset_loop/main/add_50_i15
                          reset_loop/main/add_50_i12
                          reset_loop/main/add_50_i10
                          reset_loop/main/add_50_i8
                          reset_loop/main/mul_49_i7
                          reset_loop/main/add_50_i11
                          reset_loop/main/add_50_i9
                          reset_loop/main/mul_49_i8
                          reset_loop/main/mul_45
                          reset_loop/main/mul_49_i15
                          reset_loop/main/mul_49_i14
                          reset_loop/main/mul_49_i13
                          reset_loop/main/mul_49_i11
                          reset_loop/main/mul_49_i10
                          reset_loop/main/mul_49_i9
s_2_2    -      s_2_3     reset_loop/main/result_54 (write)
                          reset_loop/main/add_50_i2
```

```
                                  reset_loop/main/add_50_i5
                                  reset_loop/main/add_50_i4
                                  reset_loop/main/add_50_i3
                                  reset_loop/main/add_50_i1
                                  reset_loop/main/add_50_i6
                                  reset_loop/main/add_50_i7
                                  reset_loop/main/mul_49_i6
                                  reset_loop/main/add_50_i0
                                  reset_loop/main/mul_49_i0
                                  reset_loop/main/mul_49_i5
                                  reset_loop/main/mul_49_i4
                                  reset_loop/main/mul_49_i3
                                  reset_loop/main/mul_49_i1
s_2_3        -       s_2_2        reset_loop/main/sample_44 (read)
                                  reset_loop/main/add_50_i13
                                  reset_loop/main/add_50_i14
                                  reset_loop/main/add_50_i15
                                  reset_loop/main/add_50_i12
                                  reset_loop/main/add_50_i10
                                  reset_loop/main/add_50_i8
                                  reset_loop/main/mul_49_i7
                                  reset_loop/main/add_50_i11
                                  reset_loop/main/add_50_i9
                                  reset_loop/main/mul_49_i8
                                  reset_loop/main/mul_45
                                  reset_loop/main/mul_49_i15
                                  reset_loop/main/mul_49_i14
                                  reset_loop/main/mul_49_i13
                                  reset_loop/main/mul_49_i11
                                  reset_loop/main/mul_49_i10
                                  reset_loop/main/mul_49_i9
```

--------------------------------------------------------------------

## 7.3   Simulation

The usual practice is to simulate the input design before you synthesize it; this lets you work out the bugs before you invest too much effort in synthesis. This chapter doesn't follow that flow, largely because the Verilog simulator doesn't allow me to use parameters the way I did above; thus my behavioral simulation of the Verilog source was necessarily of a somewhat different design, which used a memory instead of the parameter array. The simulatable design was markedly inferior after synthesis. Rather than accept a bad design, I first got it simulating correctly, then modified it (carefully) to get the version shown above; then synthesized it, and wrote out an RTL description in Verilog. The RTL version is simulatable, because the parameter array has just become a collection of hardwired mux and operator inputs. It is to be hoped that you won't often have to do this kind of thing, but it is useful at times.

One thing you need to be aware of in this context is that the RTL Verilog will have a function call that has no corresponding function definition. This comes about because we used the **map_to_operator** directive in the twos-complement multiplier that is needed to make Verilog do signed arithmetic on bit vectors. When BC encountered the **map_to_operator**, it replaced the function body with the Design-Ware operator.

When the RTL Verilog was written out, the original (simulatable) function definition was no longer present. Thus the RTL Verilog generator was left with nothing to write. The design has, after all, not been compiled to the gate level, so there are no gates in the DesignWare multiplier. So the Verilog generator writes out a call to a function **MULT_TC_OP**, but no corresponding definition of the function. You must supply the function definition. Fortunately, you already have it: it's in the source HDL. You just have to make a copy and rename it (alternatively, rename the function calls to match the existing function).

This issue does not come up if you compile the design before you write it out: then there will be gates instead of DesignWare operators and everything you need will be written out automatically.

Let's begin with the VHDL testbench.

```
library IEEE;
use IEEE.std_logic_1164.all;
use ieee.std_logic_arith.all;
entity tb_e is
end tb_e;

architecture tb_a of tb_e is
  signal clock, reset: std_logic;
  signal instream: signed (7 downto 0);
  signal outstream: signed (9 downto 0);
  component fir -- declare the filter
    port (clk, reset: in std_logic;
          sample: in signed (7 downto 0);
          result: out signed (9 downto 0));
  end component;
  begin
    filter: fir  -- instantiate the filter
      port map(  -- port    signal
                 clk    => clock,
                 reset  => reset,
                 sample => instream,
                 result => outstream);
```

```
clockgen: process begin
   clock <= '1';
   loop
      wait for 25 ns;
      clock <= not clock;
   end loop;
end process clockgen;

po_reset: process begin
   reset <= '1';
   wait for 102 ns; -- two clock cycles
   -- note the extra 2 ns to prevent races
   reset <= '0';
   wait;
end process po_reset;

stimulus: process begin
-- let's just give it a step
   instream <=  "00000000";
   wait for 302 ns;
   -- note the extra 2 ns to prevent races
   instream <=  "00010000";
   wait;
end process stimulus;
end tb_a;

configuration con of tb_e is
   for tb_a
      -- simulate the input HDL
      for filter: fir use entity work.fir(beh); end for;
      -- simulate the result at RTL
      -- for filter: fir use entity work.fir(SYN_beh); end for;
   end for;
end con;
```

The VHDL testbench consists of an entity, an architecture for the entity, and a configuration for the entity. Because the design is scheduled in superstate mode, it might have been scheduled in such a way that its timing after synthesis was not the same as it was before synthesis; in that case there would be no point to simulating the pre- and post-synthesis designs side by side. This test bench instantiates a single component, named **filter**, and three processes to drive the filter.

The first process is the clock generator. It initializes the clock signal to '1' and then inverts its polarity every 25 nanoseconds; this keeps on happening forever.

The second process generates a power-on reset pulse of two cycles' duration. Notice that in the case where pre- and post-synthesis designs are simulated side by side, this process would be too simple. An example of a more complex power-on reset process will be given in the next chapter, which demonstrates fixed I/O mode.

The third process generates the input step function. First it sets the input stream to zero, then it waits for six clock cycles, and finally it sets the input stream to hexadecimal 10. The reason to use a step function as input, instead of e.g. a unit impulse, is to avoid a situation in which the response of the pre- and post-synthesis designs would be different. As long as you don't count clock cycles, the two output sequences should be the same; this would not necessarily be the case if I had used a spike input, because the duration or time origin of the spike might have had to be adjusted. That isn't always easy: you have to figure out when the read operation is occurring and set the input during that cycle, and then reset it when you are sure it has been read.

The Verilog testbench is a little more concise. I have used an impulse function as input, but otherwise it has about the same functionality as the VHDL testbench.

Recall, however, that Verilog simulators will have problems with indexing into the parameter coefficient array in the source HDL shown above; you will have to modify the source HDL a little (or use **ugly.v** from the diskette) to simulate it. Specifically, replace the parameter with a memory, then initialize the memory, then access the memory using the index variable **i** as the address. You won't have any such trouble simulating the RTL output.

```verilog
`include "fir.v"
module tb ();

  reg clock, reset;
  reg [7:0] instream;
  wire [10:0] outstream;

  fir filter (clock, reset, instream, outstream);

  initial begin: po_reset
    reset = 1'b1;
    reset = #102 1'b0;
  end

  always begin: clockgen
    clock = 1'b1;
    forever begin
      #25 clock = ~clock;
    end
  end
```

```
initial begin: stimulus // impulse function
   #2; // skew wrt clock
   instream = 8'h0;
   #100; // after the reset
   instream = 8'h80;
   #50;
   instream = #102 8'h0;
end

always begin: monitor
   @(posedge clock);
   @(posedge clock);
   $display("time %d, input %x, output %x",
            $time, instream, outstream);
end

always begin: kill
   @(posedge clock);
   if ($time >= 2000) $stop;
end
endmodule
```

## 7.4  Decreasing cost

In order to reduce the cost of the design, we can increase the latency and/or lengthen the clock cycle. Notice that these don't always reduce cost; you have to suit the latency to the design, and sometimes try out different parameter combinations. For example, in this test case we could increase the clock cycle, but that might just increase the amount of chaining; so the amount of hardware might well increase, if we did not take measures to increase or hold constant the latency.

In order to increase the latency of a design that is scheduled in superstate mode, it is usually only necessary to add manual constraints that stretch out the time interval of interest. This interval should be that which has the greatest number of operations occurring concurrently; stretching out an interval that has few operations is unlikely to reduce cost.

Here is a modified script.

```
/* analyze -f vhdl fir.vhd */
analyze -f verilog fir.v
elaborate -s fir
create_clock clk -p 50
set_cycles 6 -from_begin fir_main/reset_loop/main \
            -to_end fir_main/reset_loop/main
```

```
/* this next is just good practice */
/* the result_54 write is only for VHDL.
   The Verilog name differs because I used a line label */
/*
set_cycles 1 -from fir_main/reset_loop/main/result_54 \
             -to_end fir_main/reset_loop/main
*/
set_cycles 1 -from fir_main/reset_loop/main/result_aut \
             -to_end fir_main/reset_loop/main
schedule -io super
report_schedule -summary        > slo_sum
report_schedule -op -mask rwo > slo_rwo
```

The first of my manual constraints forces the time from the beginning to the end of the main loop to be exactly 6 cycles. Remember that the innermost loop is being unrolled by default; so its operations will be distributed out across the latency of the next loop out, i.e. `main`.

The second constraint pulls the output write as close as it can get to the end of the loop (recall that under the superstate rules we can't push the write all the way to the bottom of the loop). This constraint is not always needed, but it is a good idea: it decreases the size of BC's search space. BC would otherwise be free to set the sizes of the loop's two superstates in any combination adding up to six, and it would have to explore all of those options to find the optimum. But it seems clear that no solution having more than one cycle in the second superstate can be better than a solution with one cycle in that superstate; the constraint forces BC to consider only solutions with a one-cycle second superstate.

Here is the summary report for the design as synthesized by the script above.

```
*******************************************
*  Summary report for process fir_main:  *
*******************************************
-------------------------------------------------------------------
      Timing Summary
-------------------------------------------------------------------
Clock period 50.00
Loop timing information:
    fir_main...........................7 cycles (cycles 0 - 7)
        reset_loop.....................7 cycles (cycles 0 - 7)
            main.......................6 cycles (cycles 1 - 7)
-------------------------------------------------------------------
      Area Summary
-------------------------------------------------------------------
Estimated combinational area    3487
Estimated sequential area       2010
TOTAL                           5497
```

```
8 control states
8 basic transitions
2 control inputs
10 control outputs


------------------------------------------------------------------

    Resource types
------------------------------------------------------------------

    Register Types
==========================================
    8-bit register.....................2
    17-bit register...................1
    19-bit register...................1
    136-bit register..................1


    Operator Types
==========================================
    (8_8->14)-bit DW02_mult............1
    (8_8->15)-bit DW02_mult............1
    (9_8->17)-bit DW02_mult............1
    (19_19->19)-bit DW01_add..........4


    I/O Ports
==========================================
    8-bit input port..................1
    10-bit registered output port......1


------------------------------------------------------------------
```

Notice that we have a third as many multipliers and half as many adders as in the first design; but the design is still more than half as large as the first design, while performing at one-third the rate. There are two main reasons for this.

First, comparing numbers of multipliers is misleading. Close examination of the multipliers in question shows us that in the faster design the multipliers are only as large as they need to be to perform a small number of operations each. For some of the multipliers, the operations were all on small operands; hence the multipliers could be small. In the slower design there are more operations on each multiplier, *including large ones*, and so the multipliers are larger. In other words, we conserved only the small multipliers.

The second reason this design costs more than half as much is that there is a large amount of multiplexing hardware that needs to be present to make the more serial design work.

Below is the operation table of the slower design. Notice that the multipliers and adders are all rather large. Notice also that we could probably eliminate an adder (**r48** or **r49**) by adding a single cycle to the schedule, and possibly we could eliminate a cycle without penalty: there appear to be just enough gaps in the schedule for the adders. However, it may be more difficult to meet our clock cycle constraint, because we might be increasing the length of the longest operation chain if we do so.

```
*********************************************
*  Operation schedule of process fir_main:  *
*********************************************

    Resource types
====================================
    p0.......8-bit input port sample
    p1.......10-bit registered output port result
    r27......(19_19->19)-bit DW01_add
    r39......(9_8->17)-bit DW02_mult
    r42......(8_8->15)-bit DW02_mult
    r43......(8_8->14)-bit DW02_mult
    r47......(19_19->19)-bit DW01_add
    r48......(19_19->19)-bit DW01_add
    r49......(19_19->19)-bit DW01_add
 c
 y    p                              m      m      m     p
 c    o    a      a      a      a    u      u      u     o
 l    r    d      d      d      d    l      l      l     r
 e    t    d      d      d      d    t      t      t     t
---+-----+------+------+------+------+------+------+------+------
   | p0  | r27  | r47  | r48  | r49  | r39  | r42  | r43  | p1
-------------------------------------------------------------------
0 |.....|......|......|......|......|......|......|......|.W51.
1 |.R55.|.o65k.|......|......|......|......|.o56..|.o61c.|.....
2 |.....|.o65g.|.o65c.|.o65e.|......|.o61e.|......|.o61d.|.....
3 |.....|.o65j.|.o65b.|......|......|.o61..|.o61g.|.o61f.|.....
4 |.....|.o65a.|.o65o.|.o65h.|......|.o61b.|.o61k.|.o61h.|.....
5 |.....|.o65l.|.o65d.|......|.o65f.|.o61a.|.o61j.|.o61i.|.....
6 |.....|.o65..|.o65n.|.o65i.|.o65m.|.o61l.|.o61m.|......|.W69.
7 |.....|......|......|......|......|......|......|......|.....

    Operation name abbreviations
========================
    R55....8-bit read reset_loop/main/sample_55
    W51....10-bit write reset_loop/result_51
```

```
W69....10-bit write reset_loop/main/result_aut
o56....(8_4->12)-bit MULT_TC_OP reset_loop/main/mul_tc_56
o61....(8_9->17)-bit MULT_TC_OP reset_loop/main/mul_tc_61_i8
o65....(19_19->19)-bit ADD_UNS_OP reset_loop/main/add_65_i1
o61a...(8_9->17)-bit MULT_TC_OP reset_loop/main/mul_tc_61_i7
o61b...(8_9->17)-bit MULT_TC_OP reset_loop/main/mul_tc_61_i9
o61c...(8_4->12)-bit MULT_TC_OP reset_loop/main/mul_tc_61_i16
o61d...(8_4->12)-bit MULT_TC_OP reset_loop/main/mul_tc_61_i15
o61e...(8_5->13)-bit MULT_TC_OP reset_loop/main/mul_tc_61_i14
o61f...(8_6->14)-bit MULT_TC_OP reset_loop/main/mul_tc_61_i12
o61g...(8_7->15)-bit MULT_TC_OP reset_loop/main/mul_tc_61_i11
o61h...(8_6->14)-bit MULT_TC_OP reset_loop/main/mul_tc_61_i10
o61i...(8_6->14)-bit MULT_TC_OP reset_loop/main/mul_tc_61_i6
o61j...(8_7->15)-bit MULT_TC_OP reset_loop/main/mul_tc_61_i5
o61k...(8_6->14)-bit MULT_TC_OP reset_loop/main/mul_tc_61_i4
o61l...(8_5->13)-bit MULT_TC_OP reset_loop/main/mul_tc_61_i2
o61m...(8_4->12)-bit MULT_TC_OP reset_loop/main/mul_tc_61_i1
o65a...(19_19->19)-bit ADD_UNS_OP reset_loop/main/add_65_i8
o65b...(19_19->19)-bit ADD_UNS_OP reset_loop/main/add_65_i12
o65c...(19_19->19)-bit ADD_UNS_OP reset_loop/main/add_65_i15
o65d...(19_19->19)-bit ADD_UNS_OP reset_loop/main/add_65_i6
o65e...(19_19->19)-bit ADD_UNS_OP reset_loop/main/add_65_i14
o65f...(19_19->19)-bit ADD_UNS_OP reset_loop/main/add_65_i7
o65g...(19_19->19)-bit ADD_UNS_OP reset_loop/main/add_65_i13
o65h...(19_19->19)-bit ADD_UNS_OP reset_loop/main/add_65_i9
o65i...(19_19->19)-bit ADD_UNS_OP reset_loop/main/add_65_i2
o65j...(19_19->19)-bit ADD_UNS_OP reset_loop/main/add_65_i11
o65k...(19_19->19)-bit ADD_UNS_OP reset_loop/main/add_65_i16
o65l...(19_19->19)-bit ADD_UNS_OP reset_loop/main/add_65_i5
o65m...(19_19->19)-bit ADD_UNS_OP reset_loop/main/add_65_i4
o65n...(19_19->19)-bit ADD_UNS_OP reset_loop/main/add_65_i3
o65o...(19_19->19)-bit ADD_UNS_OP reset_loop/main/add_65_i10
```

On the facing page is the FSM-style report on the more serial design. Notice that the number of clock ticks from the read to the write is five, and the throughput is one sample every six clock ticks.

```
present       next
state   input state    actions
--------------------------------------------------------------
s_0_0    -    s_1_1    reset_loop/result_51 (write)
s_1_1    -    s_2_2    reset_loop/main/sample_55 (read)
                       reset_loop/main/add_65_i16
                       reset_loop/main/mul_tc_56
                       reset_loop/main/mul_tc_61_i16
s_2_2    -    s_2_3    reset_loop/main/add_65_i15
                       reset_loop/main/add_65_i14
                       reset_loop/main/add_65_i13
                       reset_loop/main/mul_tc_61_i15
                       reset_loop/main/mul_tc_61_i14
s_2_3    -    s_2_4    reset_loop/main/add_65_i12
                       reset_loop/main/add_65_i11
                       reset_loop/main/mul_tc_61_i8
                       reset_loop/main/mul_tc_61_i12
                       reset_loop/main/mul_tc_61_i11
s_2_4    -    s_2_5    reset_loop/main/add_65_i8
                       reset_loop/main/add_65_i9
                       reset_loop/main/add_65_i10
                       reset_loop/main/mul_tc_61_i9
                       reset_loop/main/mul_tc_61_i10
                       reset_loop/main/mul_tc_61_i4
s_2_5    -    s_2_6    reset_loop/main/add_65_i6
                       reset_loop/main/add_65_i7
                       reset_loop/main/add_65_i5
                       reset_loop/main/mul_tc_61_i7
                       reset_loop/main/mul_tc_61_i6
                       reset_loop/main/mul_tc_61_i5
s_2_6    -    s_2_7    reset_loop/main/result_aut (write)
                       reset_loop/main/add_65_i1
                       reset_loop/main/add_65_i2
                       reset_loop/main/add_65_i4
                       reset_loop/main/add_65_i3
                       reset_loop/main/mul_tc_61_i2
                       reset_loop/main/mul_tc_61_i1
s_2_7    -    s_2_2    reset_loop/main/sample_55 (read)
                       reset_loop/main/add_65_i16
                       reset_loop/main/mul_tc_56
                       reset_loop/main/mul_tc_61_i16
--------------------------------------------------------------
```

Another thing I tried was keeping the innermost loop rolled. This was much simpler in VHDL than in Verilog, because I had to rewrite the Verilog to use a memory instead of the parameter. Keeping the innermost loop rolled results in a much longer latency, because every pass through the loop calls for at least one clock edge, and each pass will only compute the product of one coefficient. Thus we will be needing a latency of at least 17 cycles to perform this using a straightforwardly unrolled loop. I tried this, and got a total estimated area of 3566 gates. It's hard to see how to get the design into much less area without a major rethinking of the arithmetic operations, and the latency penalty is pretty severe compared to the foregoing solutions.

Another thing that could be done with this design would be to partially unroll the innermost loop. For example, imagine that we performed two coefficient multiplications in each pass through the loop; then we would need only eight passes. This is a major code rewrite but might be worth it.

Pipelining the larger multiplication operations may give us a cost per throughput benefit in the case where we are trying for a longer latency. It will also help us with the cycle time, because the multiplications will in effect be split into two or more stages of roughly equal delay. But notice that this will also add at least one cycle of latency while we wait for the pipelined components. Thus we can expect to see the best results from this approach in the region of increased latencies and shorter clock cycles. We would also expect to see increased register costs.

Pipelining the innermost loop won't help much in a case like this one, because it is best left unrolled anyway. If we left it rolled and then pipelined it, we would have dead (flushing) cycles after its exit in any case, and the dead cycles would be longer than the latency of the unrolled loop.

Pipelining the main loop (while leaving the innermost loop unrolled) will help us in the case where we are trying to achieve a single-cycle sample rate and a fast cycle time. Then loop pipelining will help because we will be able to perform all of the operations without chaining them. However, the register cost may increase.

## Summary

This chapter has walked you through a straightforward synthesis problem using BC. It gives a number of concrete and complete examples of concepts that were introduced in more fragmentary and abstract forms in previous chapters.

This chapter has also demonstrated the use of superstate mode to generate a spectrum of design implementations from a single HDL source file. The ways in which the various reports can be interpreted to answer specific questions were described as well.

You should be coming to an appreciation that the number of possible implementations of a design can be quite large. You can explore these implementations most efficiently by an intelligent combination of constraints and HDL coding techniques.

# Chapter 8

# IIR filter: handshaking I/O protocol

This chapter describes another DSP example: an infinite-impulse response (IIR) filter. This filter illustrates the basics of scheduling in fixed I/O mode. It also illustrates some techniques for writing HDL code for more complex kinds of control flow than the previous example. In particular, this filter loads a set of coefficients immediately after a reset, then executes the filtering function with two-wire handshaking synchronization. The exploration of this design includes the use of multicycle and pipelined multipliers to speed up the clock, as well as latency reduction.

The salient points of this IIR filter are: it has a simple mathematical function; it uses handshaking to control its normal-mode I/O stream; and it can easily be generalized to handle a large number of common DSP cases. The function computed by this filter is expressed by the difference equation

$$y[k] = a_1 y[k-1] + a_2 y[k-2] + b_0 u[k] + b_1 u[k-1] + b_2 u[k-2]$$

Here $y[k]$ is the $k^{th}$ output sample, and $u[k]$ is the $k^{th}$ input sample; $u[k-1]$ and $y[k-1]$ are the input and output of one sample cycle ago, and so on. This is a member of a common class of digital filters; it can be modified to provide a wide range of filter characteristics merely by setting the constant coefficients $a_1, a_2, b_0, b_1...$ appropriately. It is also easy to generalize: one can add coefficients and storage for older values of $y$ and $u$, or even eliminate the old $y$ samples entirely, and so construct a wide variety of filters.

## 8.1 Initial Design

This design has two high-level states, each of which has several machine states. In the first high-level state, the filter reads in a set of coefficients; these are the $a_i$ and the $b_i$ that describe the behavior of the filter. From the first high-level state we always go to the second; and the second is an infinite loop that first reads in a sample, then writes out a sample. A reset sends it back to the first high-level state to load a new set of coefficients.

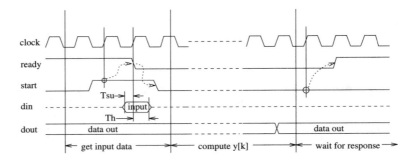

**Figure 8.1.** Timing diagram for the circuit in filter mode

I designed this filter to have loadable coefficients because I wanted to illustrate a somewhat more complicated flow of control than that of the FIR filter example of the previous chapter. One reason to build a filter like this might be that it is to be a peripheral attached to a microprocessor bus, where the microprocessor would be responsible for computing and loading in the coefficients as part of some control algorithm. This might be useful if we wanted good filter performance but didn't want the power dissipation associated with a sufficiently fast microprocessor to execute the filter algorithm directly.

A C coded version of this filter might look a little like this.

```
for (;;) {
    a1 = read(params); /* reset tail. download parameters */
    a2 = read(params);
    b0 = read(params);
    b1 = read(params);
    b2 = read(params);
    yk1 = 0; yk2 = 0; uk1 = 0; uk2 = 0;
    /* main loop */
    while (not reset) {
        u = read(din); /* handshaking is hidden! */
        yk = a1 * yk1 + a2 * yk2 +
            b0 * u + b1 * uk1 + b2 * uk2;
        write(dout, yk); /* handshaking is hidden */
        yk2 = yk1; yk1 = yk;
        uk2 = uk1; uk1 = u;
    }
}
```

The only other wrinkle here is that the sample read and write use a two-wire handshaking protocol, as shown in Fig. 8.1. **Ready** is the output that signals that the filter is ready for another sample; **start** is the input that tells the filter that it can strobe the sample and begin processing.

Here is the VHDL source text. I don't recommend that you use it as it stands in your next design: for one thing, the feedback variable can overflow on some inputs.

```vhdl
library ieee;
use ieee.std_logic_1164.all, ieee.std_logic_arith.all;
library synopsys;
use synopsys.attributes.all;
entity iir is
   port ( clk, reset, start : in  std_logic;
          din    : in signed(7 downto 0);
          params : in signed(15 downto 0);
          dout   : out signed(7 downto 0);
          ready  : out std_logic
        );
end iir;
architecture bev of iir is
begin
  main : process
    variable i: integer;
    variable uk, uk1, uk2: signed(7 downto 0);
    variable a1, a2, b0, b1, b2, yk1, yk2 : signed(15 downto 0);
    variable utmp: signed (22 downto 0);
    variable b0t, b1t, b2t: signed (23 downto 0);
    variable yk, ytmp: signed (26 downto 0);
    variable ysum: signed (28 downto 0);
    variable yo1, yo2: signed (31 downto 0);
    begin
        reset_loop: loop
            uk1 := (others => '0');
            uk2 := (others => '0');
            yk1 := (others => '0');
            yk2 := (others => '0');
            ready <= '0';
            dout <= (others => '0');
            wait until clk'event and clk='1';
            if (reset = '1') then exit reset_loop; end if;
            a1  := params;
            wait until clk'event and clk='1';
            if (reset = '1') then exit reset_loop; end if;
            a2  := params;
            wait until clk'event and clk='1';
            if (reset = '1') then exit reset_loop; end if;
            b0  := params;
            wait until clk'event and clk='1';
```

```
if (reset = '1') then exit reset_loop; end if;
b1   := params;
wait until clk'event and clk='1';
if (reset = '1') then exit reset_loop; end if;
b2   := params;
ready <= '1';
wait until clk'event and clk='1';
if (reset = '1') then exit reset_loop; end if;
outer: loop
    while (start /= '1') loop
        wait until clk'event and clk='1';
        if (reset = '1') then exit reset_loop; end if;
    end loop;
    wait until clk'event and clk='1';
    if (reset = '1') then exit reset_loop; end if;
    uk := din;
    ready <= '0';
    yo1 := yk1 * a1;
    yo2 := yk2 * a2;
    ysum := yo1(31 downto 3) + yo2(31 downto 3);
    ytmp := ysum(26 downto 0);
    b0t := uk * b0;
    b1t := uk1 * b1;
    b2t := uk2 * b2;
    utmp := b0t(22 downto 0) + b1t(22 downto 0)
            + b2t(22 downto 0);
    yk := ytmp + utmp;
    uk2 := uk1;
    uk1 := uk;
    yk2 := yk1;
    yk1 := yk(26 downto 11);
    for i in 0 to 3 loop
        wait until clk'event and clk = '1' ;
        if (reset = '1') then exit reset_loop; end if;
    end loop;
    dout <= yk(26 downto 19);
    wait until clk'event and clk = '1' ;
    if (reset = '1') then exit reset_loop; end if;
    while (start /= '0' ) loop
        wait until clk'event and clk='1';
        if (reset = '1') then exit reset_loop; end if;
    end loop;
    wait until clk'event and clk='1';
    if (reset = '1') then exit reset_loop; end if;
```

```
                    ready <= '1';
                    wait until clk'event and clk = '1' ;
                    if (reset = '1') then exit reset_loop; end if;
                end loop outer;
                wait until clk'event and clk='1';
                if (reset = '1') then exit reset_loop; end if;
            end loop reset_loop;
        end process main;
end bev;
```

Here is the Verilog text for the filter. Once again, we have to make a few changes due to Verilog's lack of twos-complement arithmetic. We also have to make a few minor changes to make the filter simulatable.

```
'define wait                                    \
  begin                                         \
    @(posedge clk);                             \
    if (reset == 1'b1) disable main;            \
  end

module iir (clk, reset, start, din, params, dout, ready);
  input clk, reset, start;
  input [7:0] din;
  input [15:0] params;
  output [7:0] dout;
  reg [7:0] dout;
  output ready;
  reg ready;

  function [31:0] mul_tc_16_16;
      input [15:0] A;
      input [15:0] B;
      reg sgn;
        // synopsys map_to_operator MULT_TC_OP
        // synopsys return_port_name Z
      begin
        sgn = A[15] ^ B[15];
        if (A[15] == 1'b1) A = ~A + 1'b1;
        if (B[15] == 1'b1) B = ~B + 1'b1;
        mul_tc_16_16 = A * B;
        if (sgn == 1'b1) mul_tc_16_16 = ~mul_tc_16_16 + 1'b1;
      end
  endfunction
  function [23:0] mul_tc_8_16;
      input [7:0] A;
```

```verilog
    input [15:0] B;
    reg sgn;
       // synopsys map_to_operator MULT_TC_OP
       // synopsys return_port_name Z
    begin
       sgn = A[7] ^ B[15];
       if (A[7] == 1'b1) A = ~A + 1'b1;
       if (B[15] == 1'b1) B = ~B + 1'b1;
       mul_tc_8_16 = A * B;
       if (sgn == 1'b1) mul_tc_8_16 = ~mul_tc_8_16 + 1'b1;
    end
  endfunction

always begin: reset_loop
    reg [3:0] i;
    reg [7:0] uk, uk1, uk2;
    reg [15:0] a1, a2, b0, b1, b2, yk1, yk2;
    reg [22:0] utmp;
    reg [23:0] b0t, b1t, b2t;
    reg [26:0] yk, ytmp;
    reg [28:0] ysum;
    reg [31:0] yo1, yo2;
    forever begin:  outer
            uk1 = 8'h0;
            uk2 = 8'h0;
            yk1 = 16'h0;
            yk2 = 16'h0;
            ready <= 1'b0;
            dout <= 8'h0;
            `wait
            a1  = params;
            `wait
            a2  = params;
            `wait
            b0  = params;
            `wait
            b1  = params;
            `wait
            b2  = params;
            ready <= 1'b1;
            `wait
            forever begin: main
                while (start != 1'b1) `wait
                `wait
```

```
                uk = din;
                ready <= 1'b0;
                yo1 = mul_tc_16_16(yk1, a1);
                yo2 = mul_tc_16_16(yk2, a2);
                ysum = yo1[31:3] + yo2[31:3];
                ytmp = ysum[26:0];
                b0t = mul_tc_8_16(uk, b0);
                b1t = mul_tc_8_16(uk1, b1);
                b2t = mul_tc_8_16(uk2, b2);
                utmp = b0t[22:0] + b1t[22:0] + b2t[22:0];
                yk = ytmp + {utmp[22], utmp[22],
                            utmp[22], utmp[22], utmp};
                uk2 = uk1;
                uk1 = uk;
                yk2 = yk1;
                yk1 = yk[26:11];
                for (i = 4'h0; i < 4'h4; i = i + 4'h1) `wait
                dout <= yk[26:19];
                `wait
                while (start != 1'b0 ) `wait
                `wait
                ready <= 1'b1;
                `wait
            end
          `wait
      end
    end
endmodule
```

The first part of the design is the reset tail. In this design, the reset tail loads the coefficient registers; these are loaded through a separate port and on a fixed schedule. At the end of the reset tail, just before the normal mode loop begins, the design asserts **ready** to signal that it is ready to begin processing input samples.

The normal mode loop is subdivided into three sections. The first section is a handshaking loop that waits for an external signal **start** to go high; when this happens a sample is read and an output signal **ready** is asserted low.

The second part of the normal mode loop computes the filter function. Here also is a small **for** loop that contains only a single clock edge statement; that gives me a single control point (the terminating iteration count, here four) that allows me to vary the latency of the filter in cycle-fixed mode. At the end of this time the output is asserted.

The third part of the normal mode loop is the trailing handshake loop; it waits for the external **start** signal to go low, at which point the filter asserts **ready** high and returns to the beginning of the cycle.

## 8.2  Simulation

To simulate the design we need a test bench. The test bench consists of a clocking
process, that generates a clock signal; a reset process, that generates two staggered
resets, as described in section 5.4; a monitor process; and two instances of our filter.

The reason we want two instances of the filter is that we want to compare the
synthesized design with the simulated one; we set up the configuration of the first
to use the original, and the second to use the synthesized design. By commenting
out the assertions in the checking process we can simulate the original alone. The
simultaneous simulation of the RTL (or gate-level) synthesized design and the original HDL illustrates one very good reason to use the cycle-fixed mode: the timing
will be indentical, except in the response to resets.

```
library IEEE;
use IEEE.std_logic_1164.all;
use ieee.std_logic_arith.all;

entity bench is
end bench;

architecture bench of bench is

    signal clock : std_logic;
    signal bsreset : std_logic;
    signal pcreset : std_logic;
    signal start : std_logic;
    signal din: signed (7 downto 0);
    signal params: signed (15 downto 0);

    signal bsready: std_logic;
    signal pcready: std_logic;
    signal bsout: signed (7 downto 0);
    signal pcout: signed (7 downto 0);

    component bsiir
        port( clk : in std_logic;
        reset : in std_logic;
        start : in std_logic;
        din: in signed (7 downto 0);
        params: in signed (15 downto 0);
        dout: out signed (7 downto 0);
        ready : out std_logic);
    end component;
    begin
    initial : bsiir
```

```
                   -- port          signal
      port map(    clk       =>     clock,
                   reset     =>     bsreset,
                   start     =>     start,
                   din       =>     din,
                   params    =>     params,
                   dout      =>     bsout,
                   ready     =>     bsready );
   final   : bsiir -- use configuration to get the synthd one
                   -- port          signal
      port map(    clk       =>     clock,
                   reset     =>     pcreset,
                   start     =>     start,
                   din       =>     din,
                   params    =>     params,
                   dout      =>     pcout,
                   ready     =>     pcready );

po_reset : process
   begin
   bsreset <= '1';
   pcreset <= '1';
   wait for 102 ns;
   pcreset <= '0';
   wait until clock'event and clock = '1';
   wait for 2 ns;
   bsreset <= '0';
   wait;
 end process;

clocking : process
   begin
     clock <= '1';
     loop
        wait for 25 ns;
        clock <= not clock;
     end loop;
   end process;

vectors : process
  begin
    wait until bsreset = '0';
    start <= '0';
    din <= (others => '0');
```

```
      params <= "0010110011111001"; -- 2CF9 a1
      wait until clock'event and clock = '1';
      wait for 2 ns;
      params <= "1101010111001110"; -- D5CE a2
      wait until clock'event and clock = '1';
      wait for 2 ns;
      params <= "0001101111100110"; -- 1BE6 b0
      wait until clock'event and clock = '1';
      wait for 2 ns;
      params <= "0011011111001101"; -- 37CD b1
      wait until clock'event and clock = '1';
      wait for 2 ns;
      params <= "0001101111100110"; -- 1BE6 b2
      wait until bsready = '1';
      wait for 2 ns;
      start <= '1';
      din <= "01100000"; -- impulse input
      wait until bsready = '0';
      wait for 2 ns;
      start <= '0';
      din <= "00000000";
      loop
         wait until bsready = '1';
         wait for 2 ns;
         start <= '1';
         wait until bsready = '0';
         wait for 2 ns;
         start <= '0';
      end loop;
   end process;

eq_comparison: process
   begin
   wait until bsreset = '0';
   foo:loop
    wait until clock'event and clock = '1';
    if (bsreset = '1') then exit foo; end if;
    assert (bsready = pcready)
       report "ready comparison failed"
       severity warning;
    assert (bsout = pcout)
       report "output comparison failed"
       severity warning;
   end loop;
```

```
        end process;
end bench;

configuration con of bench is
    for bench
       for initial:bsiir use entity WORK.iir(bev); end for;
       for final:bsiir
           use entity WORK.iir_post_compile(SYN_bev);
       end for;
    end for;
end con;
```

The Verilog test bench is similar to the VHDL test bench, except that I have commented out the second (post-synthesis) filter and the associated skewed reset.

```
`include "iir.v"
/* `include "iir_rtl.v" */

module tb ();
  reg clk, reset, rrset, start;
  reg [7:0] din;
  reg [15:0] params;
  wire [7:0] outstream;
  wire [7:0] routstream;
  wire ready, rready;

  iir filter (clk, reset, start, din, params, outstream, ready);
//iir_rtl rtl(clk, rrset, start, din, params, routstream, rready);

  always begin: clkgen
     clk = 1'b1;
     forever begin
        #25 clk = ~clk;
     end
  end

  initial begin: po_reset
     reset = 1'b1;
     rrset = 1'b1;
     #100;
     rrset = #2 1'b0;
     #50;
     reset = #2 1'b0;
  end
```

```verilog
  initial begin: stimulus
     @(negedge reset);
     start = 1'b0;
     din = 8'h0;
     params = 16'h2CF9; // a1
     @(posedge clk); #2;
     params = 16'hD5CE; // a2
     @(posedge clk); #2;
     params = 16'h1BE6; // b0
     @(posedge clk); #2;
     params = 16'h37CD; // b1
     @(posedge clk); #2;
     params = 16'h1BE6; // b2
     @(posedge ready); #2;
     start = 1'b1;
     din = 8'h60; // impulse input
     @(negedge ready); #2;
     start = 1'b0;
     din = 8'h0;
     forever begin
        @(posedge ready); #2
        #2 start = 1'b1;
        @(negedge ready); #2
        #2 start = 1'b0;
     end
  end
  always begin: monitor
     #150; // get us past the reset
     @(posedge clk);
     $display("time %d, input %x, output %x, start %x ready %x",
              $time, din, outstream, start, ready);
     /* if (ready != rready || outstream != routstream) begin
          $display("comparison failed at %d ns", $time);
          $stop;
     end */
  end
  always begin: kill
     @(posedge clk);
     if ($time >= 5000) $stop;
  end
endmodule
```

## 8.3   Synthesis

The script to synthesize our example is shown here. Notice that I have used a rather long clock period; this is in part motivated by the technology (I am using `lsi_10k`, a slow technology) but also by a desire to avoid multicycle operations. In particular, the wider multiplication operations, which will have a long delay. The clock period can be decreased later on; let me first find out roughly what the characteristics of the design are.

```
analyze -f vhdl iir.vhd /* for VHDL use this line */
/* analyze -f verilog iir.v */      /* for Verilog use this one */
elaborate -s iir
create_clock -p 50 clk
schedule /* fixed i/o mode, low effort */
report_schedule -sum > iir_sum_report
report_schedule -op  > iir_ops_report
report_schedule -var > iir_var_report
report_schedule -abs > iir_fsm_report
rename_design iir iir_post_compile
/* write -f verilog -h -out iir_rtl.v */
write -f vhdl -h -out iir_rtl.vhd
```

After synthesis we get a summary report, an operation report, a storage report, and a report on the FSM. Here is the summary report. This one is derived from the Verilog source; the line numbers don't match the VHDL.

```
bc_shell> report_schedule -sum


********************************************
*  Summary report for process reset_loop:  *
********************************************
------------------------------------------------------------------
     Timing Summary
------------------------------------------------------------------
 Clock period 50.00
 Loop timing information:
     reset_loop..........................17 cycles (cycles 0 - 17)
         outer...........................17 cycles (cycles 0 - 17)
             main........................10 cycles (cycles 6 - 16)
                 loop_77.................1 cycle  (cycles 6 - 7)
                     (exit) EXIT_L77.............. (cycle 7)
                 loop_97.................1 cycle  (cycles 13 - 14)
                     (exit) EXIT_L97.............. (cycle 14)


------------------------------------------------------------------
```

```
        Area Summary
------------------------------------------------------------------
Estimated combinational area      3539
Estimated sequential area         1930
TOTAL                             5469

15 control states
17 basic transitions
2 control inputs
18 control outputs


------------------------------------------------------------------
        Resource types
------------------------------------------------------------------
        Register Types
========================================
        1-bit register....................1
        8-bit register....................3
        16-bit register...................6
        29-bit register...................2

        Operator Types
========================================
        (16_16->32)-bit DWO2_mult.........1
        (29_29->29)-bit DWO1_add..........1

        I/O Ports
========================================
        1-bit input port..................1
        1-bit registered output port......1
        8-bit input port..................1
        8-bit registered output port......1
        16-bit input port.................1
------------------------------------------------------------------
```

Notice that this design has a latency of 10 cycles; this is a little misleading because of the csteps that are absorbed into loopbacks (see Section 2.2.5). We can get a more accurate idea of the design's timing by looking at the FSM-style report, which we will do later on; for now, let's look at the latency as being 10 cycles, and we can work on improving the performance relative to what the summary report tells us.

For our ten cycles of latency we have gotten two major functional blocks, an adder of 29 bits and a multiplier of 16 bits. All of the IIR's arithmetic operations are shared on these two units. The operation report will tell us how they are utilized.

```
bc_shell> report_schedule -op

**************************************************
*  Operation schedule of process reset_loop:  *
**************************************************

          Resource types
=======================================
    din.......8-bit input port
    dout......8-bit registered output port
    loop......loop boundaries
    p0........16-bit input port params
    p1........1-bit input port start
    p2........1-bit registered output port ready
    r50.......(16_16->32)-bit DW02_mult
    r61.......(29_29->29)-bit DW01_add
```

```
                                           D
                                     D     W
                                     W     0
                                     0     2
                                     1     _
                   p      p     p    _     m     p     p
                   o      o     o    a     u     o     o
                   r      r     r    d     l     r     r
                   t      t     t    d     t     t     t
  -------+------+-----+-----+-----+------+-----+-----+------
  cycle  | loop | p0  | din | p1  | r61  | r50 | p2  | dout
  -----------------------------------------------------------
    0    |..L3..|.....|.....|.....|......|.....|.W62.|.W63..
         |..L0..|.....|.....|.....|......|.....|.....|......
    1    |......|.R65.|.....|.....|......|.....|.....|......
    2    |......|.R67.|.....|.....|......|.....|.....|......
    3    |......|.R69.|.....|.....|......|.....|.....|......
    4    |......|.R71.|.....|.....|......|.....|.....|......
    5    |......|.R73.|.....|.....|......|.....|.W74.|......
    6    |.L11..|.....|.....|.R77.|......|.o82.|.....|......
         |..L6..|.....|.....|.....|......|.....|.....|......
    7    |.L16..|.....|.....|.....|......|.....|.....|......
         |.L15..|.....|.....|.....|......|.....|.....|......
    8    |.L12..|.....|.R79.|.....|......|.o81.|.W80.|......
    9    |......|.....|.....|.....|.o83..|.o85.|.....|......
   10    |......|.....|.....|.....|.o88..|.o86.|.....|......
   11    |......|.....|.....|.....|.o88a.|.o87.|.....|......
```

```
12    |......|.....|.....|.....|.o89..|.....|.....|.W95..
13    |..L9..|.....|.....|.....|.R97.|......|.....|......|......
14    |.L14..|.....|.....|.....|......|......|.....|......
      |.L13..|.....|.....|.....|......|......|.....|......
15    |.L10..|.....|.....|.....|......|......|......|.W99.|......
16    |..L8..|.....|.....|.....|......|......|.....|......
      |..L7..|.....|.....|.....|......|......|.....|......
17    |..L5..|.....|.....|.....|......|......|.....|......
      |..L4..|.....|.....|.....|......|......|.....|......
      |..L2..|.....|.....|.....|......|......|.....|......
      |..L1..|.....|.....|.....|......|......|.....|......
```

## Operation name abbreviations
================================

```
    L0.... reset_loop_design_loop_begin
    L1.... reset_loop_design_loop_end
    L2.... reset_loop_design_loop_cont
    L3.... outer/outer_design_loop_begin
    L4.... outer/outer_design_loop_end
    L5.... outer/outer_design_loop_cont
    L6.... outer/main/main_design_loop_begin
    L7.... outer/main/main_design_loop_end
    L8.... outer/main/main_design_loop_cont
    L9.... outer/main/loop_97/loop_97_design_loop_begin
    L10... outer/main/loop_97/loop_97_design_loop_end
    L11... outer/main/loop_77/loop_77_design_loop_begin
    L12... outer/main/loop_77/loop_77_design_loop_end
    L13... outer/main/loop_97/loop_97_design_loop_cont
    L14... outer/main/loop_97/EXIT_L97
    L15... outer/main/loop_77/loop_77_design_loop_cont
    L16... outer/main/loop_77/EXIT_L77
    R65...16-bit read outer/params_65
    R67...16-bit read outer/params_67
    R69...16-bit read outer/params_69
    R71...16-bit read outer/params_71
    R73...16-bit read outer/params_73
    R77...1-bit read outer/main/loop_77/start_77
    R79...8-bit read outer/main/din_79
    R97...1-bit read outer/main/loop_97/start_97
    W62...1-bit write outer/ready_62
    W63...8-bit write outer/dout_63
    W74...1-bit write outer/ready_74
    W80...1-bit write outer/main/ready_80
    W95...8-bit write outer/main/dout_95
```

```
    W99...1-bit write outer/main/ready_99
    o81...(16_16->32)-bit MULT_TC_OP outer/main/mul_tc_16_16_81
    o82...(16_16->32)-bit MULT_TC_OP outer/main/mul_tc_16_16_82
    o83...(29_29->29)-bit ADD_UNS_OP outer/main/add_83
    o85...(8_16->24)-bit MULT_TC_OP outer/main/mul_tc_8_16_85
    o86...(8_16->24)-bit MULT_TC_OP outer/main/mul_tc_8_16_86
    o87...(8_16->24)-bit MULT_TC_OP outer/main/mul_tc_8_16_87
    o88...(23_23->23)-bit ADD_UNS_OP outer/main/add_88_2
    o89...(27_27->27)-bit ADD_UNS_OP outer/main/add_89
    o88a..(23_23->23)-bit ADD_UNS_OP outer/main/add_88
************************************************************
```

The operation report (which I have edited a little to make it fit into a 65-column format) can be dissected to tell us that the time from reading in a sample (the operation R79 on port **din**) to writing out a result (operation W95 on port **dout**) is four csteps. That's how much time the calculation takes; the rest is overhead due to the handshaking. Don't give up in disgust, though: the overhead isn't really that bad, as we'll see in the FSM report. Between the data read R79 and the write W95 are most of the arithmetic operations o81 through o89; notice that o82 has been displaced upward to the beginning of the **main**. If we refer back to line 82 the Verilog source (that's the source of the line numbers used here) we see that o82 is the multiplication of a2 by yk2; it can be precomputed because it doesn't depend on the current sample.

The report generated by **report_schedule -var** tells us what storage we are going to pay for and what it's being used to store. I broke this one up into two variable size ranges, 16 and above and 8 and below. There are no registers between these two sizes. Here's the 16-plus report.

```
bc_shell> report_schedule -var -min 16


*******************************************
*  Register usage of process reset_loop:  *
*******************************************


            Storage resource types
=====================
    r54.......29-bit register
    r243......29-bit register
    r244......16-bit register
    r245......16-bit register
    r246......16-bit register
    r247......16-bit register
    r248......16-bit register
    r249......16-bit register
```

```
-------+------+------+------+------+------+------+------+------
cycle | r243 | r54  | r244 | r245 | r246 | r247 | r248 | r249
-----------------------------------------------------------------
      | (29) | (29) | (16) | (16) | (16) | (16) | (16) | (16)
=================================================================
   0  |......|......|......|......|......|......|......|......
   1  |......|......|..v7..|......|......|......|......|......
   2  |......|......|..v7..|..v8..|......|......|......|......
   3  |......|......|..v7..|..v8..|..v9..|......|......|......
   4  |......|......|..v7..|..v8..|..v9..|.v10..|......|......
   5  |......|......|..v7..|..v8..|..v9..|.v10..|.v11..|......
   6  |......|..v1..|.v13..|.v14..|.v15..|.v16..|.v17..|.v12..
   7  |......|..v1..|.v13..|.v14..|.v15..|.v16..|.v17..|.v12..
   8  |..v0..|..v1..|.v13..|.v14..|.v15..|.v16..|.v17..|.v12..
   9  |..v6..|..v2..|.v13..|.v14..|.v15..|.v16..|.v17..|.v12..
  10  |..v4..|..v2..|.v13..|.v14..|.v15..|.v16..|.v17..|.v12..
  11  |..v5..|..v2..|.v13..|.v14..|.v15..|.v16..|.v17..|.v12..
  12  |......|..v3..|.v13..|.v14..|.v15..|.v16..|.v17..|.v12..
  13  |......|..v3..|.v13..|.v14..|.v15..|.v16..|.v17..|.v12..
  14  |......|..v3..|.v13..|.v14..|.v15..|.v16..|.v17..|.v12..
  15  |......|..v3..|.v13..|.v14..|.v15..|.v16..|.v17..|.v12..
  16  |......|......|......|......|......|......|......|......
  17  |......|......|......|......|......|......|......|......
```

```
        Data value name abbreviations
========================
    v0.......29-bit data value outer/main/add_83/A
    v1.......29-bit data value outer/main/add_83/B
    v2.......27-bit data value outer/main/ytmp_84/var
    v3.......27-bit data value outer/main/add_89/Z
    v4.......23-bit data value outer/main/add_88_2/Z
    v5.......23-bit data value outer/main/add_88/Z
    v6.......23-bit data value outer/main/add_88_2/A
    v7.......16-bit data value outer/main/a1_loop_connect
    v8.......16-bit data value outer/main/a2_loop_connect
    v9.......16-bit data value outer/main/b0_loop_connect
    v10......16-bit data value outer/main/b1_loop_connect
    v11......16-bit data value outer/main/b2_loop_connect
    v12......16-bit data value outer/main/yk1
    v13......16-bit data value outer/main/a1
    v14......16-bit data value outer/main/a2
    v15......16-bit data value outer/main/b0
    v16......16-bit data value outer/main/b1
    v17......16-bit data value outer/main/b2
```

As you can see, the five coefficients $a_i, b_i$ take up a 16-bit register apiece; these are **r244** through **r248**. The data values for the coefficients have two names apiece; for example, the logical coefficient $a_1$ is represented by both the data value **v13** and the data value **v7**. The reason for this is the hierarchical nature of BC's scheduling representation: the same data value has different CDFG edges inside (**v13**) and outside (**v7**) the main loop. There isn't much we can do about any of the 16-bit storage if we want to keep the filter programmable.

The feedback variable **v12** is stored in the last 16-bit register; the only other big registers are the 29-bit registers holding the accumulated intermediate sums. We can probably decrease the register cost a little here, by operation reordering (rewriting the source) and/or by allowing more chaining to occur. Chaining can be increased by the use of a longer clock cycle or faster technology; or alternatively by using **set_cycles** to set the time between dependent operations to zero.

The smaller storage consists of three 8-bit registers and a 1-bit flag bit. Here is the 16-minus report that describes the small storage.

```
bc_shell> report_schedule -var -max 15


******************************************
*  Register usage of process reset_loop:  *
******************************************


            Storage resource types
=======================
      r250......8-bit register
      r302......8-bit register
      r303......8-bit register
      r312......1-bit register


-------+------+------+------+------
cycle | r250 | r303 | r302 | r312
-----------------------------------
      | (8)  | (8)  | (8)  | (1)
==================================
   0   |......|......|......|......
   1   |......|......|......|......
   2   |......|......|......|......
   3   |......|......|......|......
   4   |......|......|......|......
   5   |......|......|......|......
   6   |.v18..|.v19..|......|.v24..
   7   |.v18..|.v19..|......|......
   8   |.v18..|.v19..|.v20..|......
   9   |.v18..|.v19..|.v20..|......
```

```
10   |.v18..|.v19..|.v20..|......
11   |.v18..|......|.v20..|......
12   |.v18..|......|.v20..|......
13   |.v18..|......|.v20..|.v22..
14   |.v18..|......|.v20..|......
15   |.v18..|......|.v20..|......
16   |......|......|......|......
17   |......|......|......|......
```

```
          Data value name abbreviations
========================
     v18......8-bit data value outer/main/uk1
     v19......8-bit data value outer/main/uk2
     v20......8-bit data value outer/main/din_79/net
     v22......1-bit data value outer/main/loop_97/U2/Z
     v24......1-bit data value outer/main/loop_77/U2/Z
```

As you can see, the three 8-bit registers are used to store input samples. It's possible that scheduling operations on **uk** differently might have saved us one of these registers; but the other two are saving old values of the samples, and there's nothing we can do about them. So we might be able to save 8 bits here by forcing the first coefficient multiplication into the same step as the sample read, but even that isn't certain.

The last bit of storage we see here is a flag bit. We know this by the name of the data values: they are the results of a pair of logical operations (in different loops) both named **U2**. If we looked closely at the elaborated HDL using Design Analyzer, we would see that there are two gates named **U2**; they are the status bits that tell us to leave the handshaking loops **loop_97** and **loop_77**. BC needs to save these bits in order to make a control state branch decision; otherwise there would be a risk that the control FSM would become part of a combinational feedback cycle (see Section 4.1.1).

The next report we should look at is the FSM-style report. This report will help us to arrive at a detailed understanding of the exact timing and of the conditions under which operations and state branches will occur.

The state graph can be constructed directly from this report. The reset tail is represented by the chain of states beginning with state **s_0_0**, which is the reset state, and ending at **s_1_6**. Notice that most of what happens in the reset tail is read operations on the port **params**.

State **s_3_7** is the first state of the main loop and also of the first handshaking loop. Notice that the transition from **s_3_7** to itself occurs on input 0, while on input 1 the next state is **s_2_9**, with a read of **din** and and a write to **ready**.

```
bc_shell> report_schedule -abs

*******************************************************
*   State graph style report for process reset_loop:  *
*******************************************************
present          next
state   input    state        actions
-------------------------------------------------------
s_0_0     -      s_1_1        outer/ready_62 (write)
                              outer/dout_63 (write)
s_1_1     -      s_1_2        outer/params_65 (read)
s_1_2     -      s_1_3        outer/params_67 (read)
s_1_3     -      s_1_4        outer/params_69 (read)
s_1_4     -      s_1_5        outer/params_71 (read)
s_1_5     -      s_1_6        outer/params_73 (read)
                              outer/ready_74 (write)
s_1_6     -      s_3_7        outer/main/loop_77/start_77 (read)
                              outer/main/mul_tc_16_16_82
s_2_9     -      s_2_10       outer/main/add_83
                              outer/main/mul_tc_8_16_85
s_2_10    -      s_2_11       outer/main/add_88_2
                              outer/main/mul_tc_8_16_86
s_2_11    -      s_2_12       outer/main/mul_tc_8_16_87
                              outer/main/add_88
s_2_12    -      s_2_13       outer/main/dout_95 (write)
                              outer/main/add_89
s_2_13    -      s_4_14       outer/main/loop_97/start_97 (read)
s_2_16    -      s_3_7        outer/main/loop_77/start_77 (read)
                              outer/main/mul_tc_16_16_82
s_3_7     1      s_2_9        outer/main/din_79 (read)
                              outer/main/ready_80 (write)
                              outer/main/mul_tc_16_16_81
s_3_7     0      s_3_7        outer/main/loop_77/start_77 (read)
s_4_14    1      s_2_16       outer/main/ready_99 (write)
s_4_14    0      s_4_14       outer/main/loop_97/start_97 (read)
-------------------------------------------------------
         **********         Branch Conditions        **********
-------------------------------------------------------
state            condition        source
-------------------------------------------------------
s_3_7            1                2nd branch of outer/main/loop_77/SPLIT_L77
s_3_7            0                1st branch of outer/main/loop_77/SPLIT_L77
s_4_14           1                2nd branch of outer/main/loop_97/SPLIT_L97
s_4_14           0                1st branch of outer/main/loop_97/SPLIT_L97
```

From state **s_2_9** to state **s_2_13** the arithmetic is performed. This is a total of four states, with one extra if you count the transition from **s_3_7** to **s_2_9**. The transition from **s_2_12** to state **s_2_13** is where the result write occurs.

From **s_2_13** the filter goes to **s_4_14**, which is the second handshaking loop. This loop is exited on a 1, whence it goes to **s_2_16**, and thence back to **s_3_7** where the cycle begins again. Counting transitions and assuming that the external logic can respond immediately, we see that the shortest sample cycle time is 8 cycles.

### Trimming latency

The first thing we can do to speed up this design is eliminate the extra clock cycle associated with state **s_2_16**. This is here because of the extra clock edge statement at the bottom of the process in the source HDL. Note that this extra edge isn't really apparent in any of the reports; sometimes you just have to look at your source code to see optimizations.

I had put that clock edge in originally because I wrote the test case with superstate I/O mode in mind. When I converted the design to fixed I/O mode, the extra clock edge became redundant. if we take out this clock edge, the design will lose one cycle of latency with almost no area penalty. This is a common source of needless latency: any time you move a design from superstate mode to fixed mode you should look for 'extra' clock edges like this one.

The next thing we can do is reduce the number of cycles of delay in the main loop. Recall that there were four cycles in the **for** loop of the original specification; I set this number to two and ran the scheduler again, with the result that the total estimated cost went from 5469 to 7193, but the fastest possible sample cycle is now only 5 cycles. Here's the relevant part of the FSM report. Notice that a number of operations have been shuffled out to the ends of the loop on the transitions from **s_4_12** to **s_3_7** and from **s_3_7** to **s_2_9**. Of course this speedup has cost me something: I now have two adders and two multipliers. I have marked the sample read and write operations with strings of asterisks, and reordered the table a bit.

```
bc_shell> report_schedule -abs
```

| cur<br>state | cond | next<br>state | actions |
|---|---|---|---|
| s_3_7 | 0 | s_3_7 | outer/main/loop_77/start_77 (read) |
| s_3_7 | 1 | s_2_9 | outer/main/din_79 (read) ***********<br>outer/main/ready_80 (write)<br>outer/main/add_88_2<br>outer/main/mul_tc_8_16_87<br>outer/main/mul_tc_8_16_85 |
| s_2_9 | - | s_2_10 | outer/main/add_88<br>outer/main/mul_tc_16_16_81 |
| s_2_10 | - | s_2_11 | outer/main/dout_95 (write) ********* |

```
                         outer/main/add_83
                         outer/main/add_89
s_2_11   -      s_4_12   outer/main/loop_97/start_97 (read)
s_4_12   1      s_3_7    outer/main/loop_77/start_77 (read)
                         outer/main/ready_99 (write)
                         outer/main/mul_tc_16_16_82
                         outer/main/mul_tc_8_16_86
s_4_12   0      s_4_12   outer/main/loop_97/start_97 (read)
```

## 8.4  Speeding up the Clock

Now let's try to decrease the clock period. We can start by looking at the delay report; it appears that the multiplications are the real bottleneck.

```
Cumulative delay starting at mul_tc_16_16_81:
  mul_tc_16_16_81  =  45.855202
    add_89  =  60.178314
    dout_95  =  60.178314
  add_83  =  60.407917
```

The first thing we can try is the use of a multicycle multiplier. To get this, all we need to do is reduce the clock period; the multiplier will be scheduled in two or more cycles automatically. However, this will cost us considerable latency, something we'd rather not give up unnecessarily. So instead, I will take the design with a five-state minimum sample cycle, and tell BC to use two-stage pipelined multipliers. This will add some latency, because of the latency of the multipliers; but perhaps I can cut down the clock cycle and so get back the time I sacrifice.

To do this I have to include a new line in my synthesis script:

synthetic_library = {dw01.sldb dw02.sldb dw03.sldb}

I also had to modify the source HDL. Here is the Verilog to use a pipelined multiplier. The difference is that the multiplications are mapped to a different multiplier, and the parameter CLK has been added to the multiplication function. Notice that the parameter CLK must be spelled using only upper-case characters: this is a requirement for all sequential components.

```
'define wait                                   \
   begin @(posedge clk);                       \
   if (reset == 1'b1) disable reset_loop;      \
   end

module iir (clk, reset, start, din, params, dout, ready);
  input clk, reset, start;
  input [7:0] din;
  input [15:0] params;
```

```verilog
output [7:0] dout;
reg [7:0] dout;
output ready;
reg ready;

function [31:0] mul_tc_16_16;
    input [15:0] A;
    input [15:0] B;
    input CLK;
    reg sgn;
      // commented out synopsys map_to_operator MULT_TC_OP
      // synopsys map_to_operator DW03_mult_2_stage_TC_OP
      // synopsys return_port_name Z
    begin
      sgn = A[15] ^ B[15];
      if (A[15] == 1'b1) A = ~A + 1'b1;
      if (B[15] == 1'b1) B = ~B + 1'b1;
      mul_tc_16_16 = A * B;
      if (sgn == 1'b1) mul_tc_16_16 = ~mul_tc_16_16 + 1'b1;
    end
  endfunction

function [23:0] mul_tc_8_16;
    input [7:0] A;
    input [15:0] B;
    input CLK;
    reg sgn;
      // synopsys map_to_operator DW03_mult_2_stage_TC_OP
      // comment out synopsys map_to_operator MULT_TC_OP
      // synopsys return_port_name Z
    begin
      sgn = A[7] ^ B[15];
      if (A[7] == 1'b1) A = ~A + 1'b1;
      if (B[15] == 1'b1) B = ~B + 1'b1;
      mul_tc_8_16 = A * B;
      if (sgn == 1'b1) mul_tc_8_16 = ~mul_tc_8_16 + 1'b1;
    end
  endfunction

always begin: reset_loop
    reg [3:0] i;
    reg [7:0] uk, uk1, uk2;
    reg [15:0] a1, a2, b0, b1, b2, yk1, yk2;
    reg [22:0] utmp;
```

```
reg [23:0] b0t, b1t, b2t;
reg [26:0] yk, ytmp;
reg [28:0] ysum;
reg [31:0] yo1, yo2;
forever begin: outer
        uk1 = 8'h0;
        uk2 = 8'h0;
        yk1 = 16'h0;
        yk2 = 16'h0;
        ready <= 1'b0;
        dout <= 8'h0;
        'wait
        a1  = params;
        'wait
        a2  = params;
        'wait
        b0  = params;
        'wait
        b1  = params;
        'wait
        b2  = params;
        ready <= 1'b1;
        'wait
        forever begin: main
            while (start != 1'b1) 'wait
            'wait
            uk = din;
            ready <= 1'b0;
            yo1 = mul_tc_16_16(yk1, a1, clk);
            yo2 = mul_tc_16_16(yk2, a2, clk);
            ysum = yo1[31:3] + yo2[31:3];
            ytmp = ysum[26:0];
            b0t = mul_tc_8_16(uk, b0, clk);
            b1t = mul_tc_8_16(uk1, b1, clk);
            b2t = mul_tc_8_16(uk2, b2, clk);
            utmp = b0t[22:0] + b1t[22:0] + b2t[22:0];
            yk = ytmp + {utmp[22], utmp[22],
                        utmp[22], utmp[22], utmp};
            uk2 = uk1;
            uk1 = uk;
            yk2 = yk1;
            yk1 = yk[26:11];
            for (i = 4'h0; i < 4'h2; i = i + 4'h1) 'wait
            dout <= yk[26:19];
```

```
                           `wait
                           while (start != 1'b0 ) `wait
                           `wait
                           ready <= 1'b1;
                    end
                    `wait
              end
        end
endmodule
```

The VHDL takes a little more modification, because it did not already have its multiplications implemented by functions. I had to add a new function, **mult_2s**, that performs multiplications when simulated, and that maps to the pipelined multiplier when run through synthesis.

```
library ieee;
use ieee.std_logic_1164.all, ieee.std_logic_arith.all;
library synopsys;
use synopsys.attributes.all;
entity iir is
   port ( clk, reset, start : in  std_logic;
           din    : in signed(7 downto 0);
           params : in signed(15 downto 0);
           dout   : out signed(7 downto 0);
           ready  : out std_logic
         );
end iir;
architecture bev of iir is
   function mult_2s(A, B : in signed ; CLK : in std_logic)
     return signed is
       begin
       -- pragma map_to_operator DW03_mult_2_stage_TC_OP
       -- pragma return_port_name Z
       return A * B;
       end;
begin
  main : process
    variable i: integer;
    variable uk, uk1, uk2: signed(7 downto 0);
    variable a1, a2, b0, b1, b2, yk1, yk2 : signed(15 downto 0);
    variable utmp: signed (22 downto 0);
    variable b0t, b1t, b2t: signed (23 downto 0);
    variable yk, ytmp: signed (26 downto 0);
    variable ysum: signed (28 downto 0);
    variable yo1, yo2: signed (31 downto 0);
```

```
begin
   reset_loop: loop
       uk1 := (others => '0');
       uk2 := (others => '0');
       yk1 := (others => '0');
       yk2 := (others => '0');
       ready <= '0';
       dout <= (others => '0');
        wait until clk'event and clk='1';
        if (reset = '1') then exit reset_loop; end if;
       a1  := params;
        wait until clk'event and clk='1';
        if (reset = '1') then exit reset_loop; end if;
       a2  := params;
        wait until clk'event and clk='1';
        if (reset = '1') then exit reset_loop; end if;
       b0  := params;
        wait until clk'event and clk='1';
        if (reset = '1') then exit reset_loop; end if;
       b1  := params;
        wait until clk'event and clk='1';
        if (reset = '1') then exit reset_loop; end if;
       b2  := params;
       ready <= '1';
        wait until clk'event and clk='1';
        if (reset = '1') then exit reset_loop; end if;
       outer: loop
           while (start /= '1') loop
              wait until clk'event and clk='1';
              if (reset = '1') then exit reset_loop; end if;
           end loop;
            wait until clk'event and clk='1';
            if (reset = '1') then exit reset_loop; end if;
           uk := din;
           ready <= '0';
           yo1 := mult_2s(yk1, a1, clk);
           yo2 := mult_2s(yk2, a2, clk);
           ysum := yo1(31 downto 3) + yo2(31 downto 3);
           ytmp := ysum(26 downto 0);
           b0t := mult_2s(uk, b0, clk);
           b1t := mult_2s(uk1, b1, clk);
           b2t := mult_2s(uk2, b2, clk);
           utmp := b0t(22 downto 0) + b1t(22 downto 0)
                   + b2t(22 downto 0);
```

```
            yk := ytmp + utmp;
            uk2 := uk1;
            uk1 := uk;
            yk2 := yk1;
            yk1 := yk(26 downto 11);
            for i in 0 to 1 loop
                wait until clk'event and clk = '1' ;
                if (reset = '1') then exit reset_loop; end if;
            end loop;
            dout <= yk(26 downto 19);
             wait until clk'event and clk = '1' ;
             if (reset = '1') then exit reset_loop; end if;
            while (start /= '0' ) loop
                wait until clk'event and clk='1';
                if (reset = '1') then exit reset_loop; end if;
            end loop;
             wait until clk'event and clk='1';
             if (reset = '1') then exit reset_loop; end if;
            ready <= '1';
          end loop outer;
           wait until clk'event and clk='1';
           if (reset = '1') then exit reset_loop; end if;
      end loop reset_loop;
    end process main;
end bev;
```

The net result of this change is an increase in area to an estimated 12129 gates. This increase in cost has bought me no increase in speed as yet: I will now decrease the clock cycle by modifying the script so that the clock is faster. A quick look at the timing report tells me I can't do much better than 28 or so nanoseconds with this multiplier, even allowing only 2 ns for the control and multiplexing delays:

```
Cumulative delay starting at mul_tc_16_16_85:
   mul_tc_16_16_85  =  26.363899
     add_93  =  38.219395
      dout_99  =  38.219395
    add_87  =  38.576996
```

So I will try a clock period of 30 ns.

```
        create_clock clk -p 30
```

The design is now overconstrained, and the attempt to schedule the design with this clock setting cannot succeed, due to the excessively long delays of the necessary operation chains. This is the error message I got:

```
Unsatisfiable Fixed Schedules found:
  (T_din_83, 2) and
  (T_dout_99, 4)
 this violates min delay of 3 cycles
   from T_din_83
   to T_dout_99
Critical Paths from T_din_83 to T_dout_99:
Path 1
din_83                     (0)     0
 add_92_2                  (1)     1
  add_92                   (0)     1
   Patch_box_26            (0)     1
    add_93                 (1)     2
     Patch_box_23_dup_22   (1)     3
      dout_99              (0)     3
Error: Fixed IO schedule is unsatisfiable (HLS-52)
```

My response to this is to increase the latency by another cycle. Since the best latency I have achieved so far is 5 cycles per sample at 50 ns per cycle, i.e. 250 ns per sample, I can afford to add another cycle at 30 ns clock period, for a net gain of 70 ns, i.e. I've increased performance about a quarter. So I set the loop that delays the main loop back to 3 states:

```
        for i in 0 to 2 loop ... end loop
```

This, however, doesn't work because of another deep chain. It appears that it is going to be difficult to achieve high performance with a 30 ns clock; so I am going to relax the clock cycle time to 35 ns. If I do so, I may be able to push the latency back down:

```
        bc_shell> create_clock clk -p 35

        for i in 0 to 1 loop ... end loop
```

This time it goes through, for a total estimated area of 16000 gates. The performance is five cycles per sample at 35 nanoseconds each, i.e. 175 nanoseconds per sample, so I have done about as well as I could have done with 30 ns clock period and six cycles of latency. As always, you should bear in mind that I'm using the **lsi_10k** technology library: presumably with a more modern library I'd be able to do much better.

The main reason for the high cost of the fastest achieved design is that it has no less than five of the pipelined multipliers; these multipliers are not shared at all, but they can all be run in parallel in two 35 ns cycles. They can't really be shared because there is no latency left in which they could deliver their outputs. Thus if we added some latency we should see a substantial decrease in cost, because the multipliers would be shareable.

Examination of the difference equation of this filter shows us that the multipliers are going to be a scarce resource no matter what we do. If their utilization was less during the critical portion of the main loop, we might have been able to make better use of the pipelining feature; but because we are trying to perform the multiplications and then sum the results over a two-cycle span, we can't expect pipelined components to do anything more than decrease the clock period. If there were another cycle of latency available this wouldn't be the case: then we could use the pipelined multipliers twice.

Loop pipelining won't help in this case, because the main loop that we would want to pipeline contains other loops that are not unrolled. Because the nested loops have data-dependent delays, pipelining this loop is not possible: there would be no predictable pattern of loop overlaps.

Here is the operation report for the 35 ns design with pipelined multipliers; it should help to make the foregoing discussion of utilization more concrete. I have edited the names of the multiplication operations to make them shorter, and I have not included most of the reads, writes, and loop operations. The data read occurs in step 8; the write occurs in step 10.

Notice that each multiplier is used to perform two operations. These are the two operations into which each multiplication has been split; that is, the $g$ and $h$ functions that compose to form a multiplication $f(x, y) = g(h(x, y))$. No more operations can be put onto these multipliers because there isn't enough time to get the results out again.

```
bc_shell> report_schedule -op

*********************************************************************
                Date      : Fri Feb 16 15:58:42 1996
                Version   : v3.4a
                Design    : iir
*********************************************************************


*********************************************
*  Operation schedule of process reset_loop:  *
*********************************************

        Resource types
===================================
    r39.......(29_29->29)-bit DW01_add
    r108......(16_16_1->32)-bit DW03_mult_2_stage
    r127......(27_27->27)-bit DW01_add
    r230......(8_16_1->24)-bit DW03_mult_2_stage
    r248......(16_16_1->32)-bit DW03_mult_2_stage
    r249......(16_8_1->24)-bit DW03_mult_2_stage
    r250......(16_8_1->24)-bit DW03_mult_2_stage
```

```
                              D      D      D      D      D
                              W      W      W      W      W
                              0      0      0      0      0
                              3      3      3      3      3
                              _      _      _      _      _
                              m      m      m      m      m
                              u      u      u      u      u
                              l      l      l      l      l
                              t      t      t      t      t
                D      D      _      _      _      _      _
                W      W      2      2      2      2      2
                0      0      _      _      _      _      _
                1      1      s      s      s      s      s
                _      _      t      t      t      t      t
                a      a      a      a      a      a      a
                d      d      g      g      g      g      g
                d      d      e      e      e      e      e
-------+-----+------+------+------+------+------+------+------
cycle  | I/O | r127 | r39  | r249 | r230 | r250 | r248 | r10
-------|-----|-------------------------------------------------
  0    |.....|......|......|......|......|......|......|......
  1    |.....|......|......|......|......|......|......|......
  2    |.....|......|......|......|......|......|......|......
  3    |.....|......|......|......|......|......|......|......
  4    |.....|......|......|......|......|......|......|......
  5    |.....|......|......|......|......|......|......|......
  6    |.....|......|......|......|......|......|......|......
  7    |.....|......|......|......|......|......|......|......
  8    |.R83.|......|......|..or..|..ot..|..ov..|..ox..|..oz..
  9    |.....|.o92..|.o92a.|..os..|..ou..|..ow..|..oy..|..oA..
 10    |.W99.|.o93..|.o87..|......|......|......|......|......
 11    |.....|......|......|......|......|......|......|......
 12    |.....|......|......|......|......|......|......|......
 13    |.....|......|......|......|......|......|......|......
```

```
R83...8-bit read outer/main/din_83
W99...8-bit write outer/main/dout_99
o87...(29_29->29)-bit ADD_UNS_OP outer/main/add_87
o92...(23_23->23)-bit ADD_UNS_OP outer/main/add_92
o93...(27_27->27)-bit ADD_UNS_OP outer/main/add_93
o92a..(23_23->23)-bit ADD_UNS_OP outer/main/add_92_2
oA....(1->32)-bit mult_state_1 outer/main/mul_tc_16_16_85
or....(8_16_1)-bit mult_state_0 outer/main/mul_tc_8_16_91
os....(1->24)-bit mult_state_1 outer/main/mul_tc_8_16_91
```

```
ot....(8_16_1)-bit mult_state_0 outer/main/mul_tc_8_16_90
ou....(1->24)-bit mult_state_1 outer/main/mul_tc_8_16_90
ov....(8_16_1)-bit mult_state_0 outer/main/mul_tc_8_16_89
ow....(1->24)-bit mult_state_1 outer/main/mul_tc_8_16_89
ox....(16_16_1)-bit mult_state_0 outer/main/mul_tc_16_16_86
oy....(1->32)-bit mult_state_1 outer/main/mul_tc_16_16_86
oz....(16_16_1)-bit mult_state_0 outer/main/mul_tc_16_16_85
```

The next thing to do is to look at the structure of the dataflow graph in the region of the multiplications. That graph looks like a tree of additions driven by leaf nodes that are multiplications. The structure of the tree is what's important: this can be derived by thinking about how the HDL will be parsed, by looking at the post-elaboration circuit in Design Analyzer, or by looking at the reports and the post-scheduling design. The conclusion we can draw is that this tree isn't always the best.

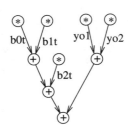

The reason the tree we have above isn't universally good is that it can't be cut up into csteps in an efficient way for building slow, cheap designs. We have five multiplications; so presumably a five-cycle design could be constructed using one multiplier. But the last multiplication would have to drive a chain of two adders: thus any five-cycle solution with one multiplier has two adders and a deep chain in one cycle.

Here's an alternative dataflow that gets us out of having to have two adders and a deep chain. In this one, notice, the last multiplication is one operation from the result; and by scheduling it as shown we can achieve five cycles with one adder, one multiplier, and chain depth of two. That's about as cheaply as it can be done without modifying the arithmetic functions.

The only problem with the 'improved' dataflow is that it won't work as well for high performance designs, i.e. those that have two or more multipliers. If we want to do that, we have to flatten out the tree as much as possible, to increase the parallelism of the design. We could do that with a three-input carry-save adder; but that could cost us some registers, depending on how it was scheduled. It might not be worth it in the end, because in order to use the three inputs all at the same time we'd have to compute three products simultaneously; that means three multipliers.

Another thing we might want to try would be a pipelined multiply-accumulate part; we can do that in BC by using DesignWare encapsulation. In the chapter 10 and in Appendix A encapsulation of combinational and sequential components using DesignWare is described.

We can still increase the performance by rewriting the source HDL. For example, there is nothing stopping us from computing $a_1 y[k]$; in the sample cycle after we compute it, the result will be exactly $a_1 y[k-1]$. That in turn will let us shorten the path from the sample input to the result output; in the most extreme case that could all be done in one cycle, because only a single multiplication and a single addition would have to be done to compute $y[k]$.

## Summary

In this chapter we have seen how to build a cycle-fixed mode design with hand-shaking I/O. We also looked at ways to improve the performance of such a design, starting with an initial design that had a lot of extra latency, and ending up with a more expensive design that had less latency and a faster clock.

We have seen the use of pipelined parts to speed up the clock cycle; and discussed various options that would stem from reorganizing the arithmetic in the source code.

# Chapter 9

# The Inverse Discrete Cosine Transform: C to HDL

In this chapter we will look at a deceptively simple problem: an inverse discrete cosine transform. The one we will look at is two-dimensional, a variation used in video and other picture-oriented compression applications.

This chapter begins with the mathematical definition of the transform. The transform is then mapped to a reasonable C language implementation. Then the C language description will be translated directly into VHDL and Verilog. The design we achieve by doing a direct translation is, however, not a very good one. We will therefore improve the HDL versions in a step-by-step process. At each stage, we will encounter some aspect of a reasonable C-language style that is not really all that good for synthesis.

The IDCT can be thought of as a pair of matrix multiplications. $O$ is the output matrix, $C$ is a coefficient matrix, $X$ is the input, and $X^T$ is the transpose of $X$.

$$O = C \times (C \times X^T)^T$$

## 9.1 Initial C Code

We will start with the assumption that the input and output are to be stored in RAM. Now a simple way to begin to think about the organization of this computation is just to write it out in a software language like C.

```
int row, col, elt, c[8][8], x[8][8], o[8][8], temp[8][8];
for (row = 0; row < 8; row++) { /* T = C X transpose(X) */
    for (col = 0; col < 8; col++) {
        temp[row][col] = 0;
        for (elt = 0; elt < 8; elt++) {
            temp[row][col] += c[row][elt] * x[col][elt];
        }
    }
}
```

```
/* compute C * transpose(temp) */
for (row = 0; row < 8; row++) {
    for (col = 0; col < 8; col++) {
        o[row][col] = 0;
        for (elt = 0; elt < 8; elt++) {
            o[row][col] += c[row][elt] * temp[col][elt];
        }
    }
}
```

This is a reasonable way to organize the computation in C; after all, the implicit model of computation that underlies C is one of serial execution and a single word-wide memory. However, hardware has more freedom.

The first problem we have in translating C directly into HDL for BC is that in a hardware implementation we are not restricted to 32-bit integers and floats. Thus we have to specify the bit widths, roundings, and truncations of the computation. For this application, the true bit width of the input elements is twelve bits, the coefficients are thirteen bits, and the output is truncated to nine bits.

Second, the serial assumption of C can be retracted in hardware. That is, we can perform many multiplications and additions in parallel. This is important, because we have 1024 multiplication operations in this transform.

Third, in C there are certain atomic operations that correspond directly to single instructions of the underlying CPU. Examples of these are integer addition and comparison. Other operations, such as cosines and logarithms, may be implemented as hardware routines on a floating-point unit, or they may be implemented in software. In synthesized hardware, by contrast, one can easily imagine constructing complex functions as single synthetic library (DesignWare) parts. For example, in the case shown here, we might use a function that would multiply and then sum several elements of a pair of vectors simultaneously using a carry-save tree.

Fourth, the serial assumption of C means that there is no very great penalty associated with accessing memory elements such as x[elt][col]. In hardware, by contrast, we can have memories with multiple ports, and in addition we can choose the bit widths of our memories to be whatever we want.

Fifth, in C there is usually an optimizing compiler that helps us by assigning operands to memory and registers. In hardware, this is a much less constrained problem, and it is correspondingly harder to construct good optimizers. Thus the user of a general-purpose tool such as BC must take control of many of the decisions about where operands are stored, how they are organized, and so on; the best way to express these decisions is in the HDL. In other words, we have to write the HDL in such a way that it expresses carefully thought out decisions about where operands are stored and how they are organized, or we will get bad designs.

Specifically, the organization implied by mapping the arrays of the C code directly into memory will get us into trouble with memory bottlenecks. There are two main ways in which we will get into trouble. First, memory access dependencies (recall the discussion of Section 3.6) will be created between the reads and writes of the RAMs. These dependencies, which may be real ones, will create scheduling constraints that won't allow us to achieve short latencies. Second, consider that a multiplication and addition may be fast compared to a RAM access. If this is the case, and if we cannot access many operands in each RAM cycle, we should expect that the RAM ports may become the bottleneck for our overall throughput for this reason as well.

## 9.2   Translation into HDL

Let's examine the effects of these considerations by gradually transforming our initial C program into VHDL and Verilog.

First, we will need a pinout, a clock, and so on. In addition, we need the apparatus necessary to load and unload the input and output RAMs; and we need to define an initial set of RAMs and variables.

Here is the VHDL for our IDCT.

```
library ieee;
use ieee.std_logic_1164.all;
use ieee.std_logic_arith.all;

package idct_types is
    type xt is array (63 downto 0) of signed (11 downto 0) ;
    type ct is array (63 downto 0) of signed (12 downto 0) ;
    type ot is array (63 downto 0) of signed (8 downto 0) ;
    type tt is array (63 downto 0) of signed (15 downto 0) ;
end idct_types;

-- package body idct_types is end idct_types;

library ieee;
use ieee.std_logic_1164.all;
use ieee.std_logic_arith.all;
library synopsys;
use synopsys.attributes.all;
use work.idct_types.all;

entity idct is
 port (clk, reset, start: in std_logic;
       done: out std_logic;
       din: in signed (11 downto 0);
       dout: out signed (8 downto 0));
end idct;
```

```
architecture beh of idct is
begin
   main: process

   variable i,j,k: signed (3 downto 0);
   variable temp25: signed (24 downto 0);
   variable temp29: signed (28 downto 0);
   variable x: xt;
   variable c: ct;
   variable o: ot;
   variable t: tt;
   begin

   -- for simulation purposes I have to load up the ROM
   -- I will use an identity matrix just for testing
   -- synopsys translate_off
      for i in 0 to 7 loop
         for j in 0 to 7 loop
            if (i = j) then
               c(8 * i + j) := ('0', '1', others => '0');
            else
               c(8 * i + j) := (others => '0');
            end if;
         end loop;
      end loop;
   -- synopsys translate_on

      reset_loop: loop
         done <= '0';
         while (start = '0') loop
            wait until clk'event and clk = '1';
            if reset = '1' then exit reset_loop; end if;
         end loop;
         wait until clk'event and clk = '1';
         if reset = '1' then exit reset_loop; end if;

         -- Load the Input RAM
         for i in 0 to 7 loop
            for j in 0 to 7 loop
               x(i * 8 + j) := din;
               wait until clk'event and clk = '1';
               if reset = '1' then exit reset_loop; end if;
            end loop;
         end loop;
```

```
        wait until clk'event and clk = '1';
        if reset = '1' then exit reset_loop; end if;
        -- first multiplication. temp = c * trans(x)
        for i in 0 to 7 loop
            for j in 0 to 7 loop
                temp25 := (others => '0');
                for k in 0 to 7 loop
                    temp25 := temp25 + (c(8 * i + k) * x(8 * k + j));
                end loop;
                t(8 * i + j) := temp25(23 downto 8);
            end loop;
        end loop;
        wait until clk'event and clk = '1';
        if reset = '1' then exit reset_loop; end if;
        -- second multiplication. o = c * trans(t)
        for i in 0 to 7 loop
            for j in 0 to 7 loop
                temp29 := (others => '0');
                for k in 0 to 7 loop
                    temp29 := temp29 + (c(8 * i + k) * t(8 * k + j));
                end loop;
                o(8 * i + j) := temp29(26 downto 18);
            end loop;
        end loop;
        wait until clk'event and clk = '1';
        if reset = '1' then exit reset_loop; end if;
        -- write out the result
        done <= '1';
        while (start = '0') loop
            wait until clk'event and clk = '1';
            if reset = '1' then exit reset_loop; end if;
        end loop;
        wait until clk'event and clk = '1';
        if reset = '1' then exit reset_loop; end if;
        for i in 0 to 7 loop
            for j in 0 to 7 loop
                dout <= o( 8 * i + j);
                wait until clk'event and clk = '1';
                if reset = '1' then exit reset_loop; end if;
            end loop;
        end loop;
    end loop;
  end process;
end beh;
```

Here is the Verilog for the same example. Notice the signed arithmetic functions `mult13_12` and `mult13_16`: these are here only for simulation. I don't need to worry about them getting synthesized because of the **map_to_operator** pragma, which causes the function bodies to be replaced.

```verilog
'define wait                                                    \
    begin                                                       \
        @(posedge clk); if (reset) disable reset_loop; \
    end

module idct(clk, reset, start, done, din, dout);
   input clk, reset, start;
   input [11:0] din;
   output done;
   reg     done;
   output [8:0] dout;
   reg     [8:0] dout;

   reg [3:0] i,j,k;
   reg [24:0] temp25;
   reg [28:0] temp29;
   reg [11:0] x [63:0];
   reg [12:0] c [63:0];
   reg [8:0]  o [63:0];
   reg [15:0] t [63:0];
   function [24:0] mult13_12;
     /* twos complement multiplication */
     input [12:0] A; // the port names A, B, Z
     input [11:0] B; // are taken from the designWare
                     // otherwise this won't work
     reg sgn;
       // synopsys map_to_operator MULT_TC_OP
       // synopsys return_port_name Z
       // the above tells how to map this whole function
       // to one designWare function.
     begin
       sgn = A[12] ^ B[11];
       if (A[12] == 1'b1) A = ~A + 1'b1;
       if (B[11] == 1'b1) B = ~B + 1'b1;
       mult13_12= A * B;
       if (sgn == 1'b1) mult13_12 = ~mult13_12 + 1'b1;
     end
   endfunction
```

```verilog
function [24:0] mult13_16;
  /* twos complement multiplication */
  input [12:0] A; // the port names A, B, Z
  input [15:0] B; // are taken from the designWare
                  // otherwise this won't work
  reg sgn;
    // synopsys map_to_operator MULT_TC_OP
    // synopsys return_port_name Z
    // the above tells how to map this whole function
    // to one designWare function.
  begin
    sgn = A[12] ^ B[15];
    if (A[12] == 1'b1) A = ~A + 1'b1;
    if (B[15] == 1'b1) B = ~B + 1'b1;
    mult13_16 = A * B;
    if (sgn == 1'b1) mult13_16 = ~mult13_16 + 1'b1;
  end
endfunction

// for simulation purposes I have to load up the ROM
// I will use an identity matrix just for testing
/* synopsys translate_off */
  initial begin: load_rom
    for (i = 0; i < 8; i = i + 1) begin
      for (j = 0; j < 8; j = j + 1) begin
        if (i == j) c[8 * i + j] = 13'h0800;
        else c[8 * i + j] = 13'h0000;
      end
    end
  end
/* synopsys translate_on */

always begin: reset_loop
  done <= 1'b0;
  while (!start) `wait `wait // loop wait plus tail wait
  // Load the Input RAM
  for (i = 0; i < 8; i = i + 1) begin
    for (j = 0; j < 8; j = j + 1) begin
      x[i * 8 + j] = din;
      `wait
    end
  end
  `wait
```

```
        // first multiplication. t = c * trans(x)
        for (i = 0; i < 8; i = i + 1) begin
            for (j = 0; j < 8; j = j + 1) begin
                t[8 * i + j] = 16'h0;
                for (k = 0; k < 8; k = k + 1) begin
                    temp25 = mult13_12(c[8 * i + k], x[8 * j + k]);
                    t[8 * i + j] = t[8 * i + j] + temp25[23:8];
                end
            end
        end
        'wait
        // second multiplication. o = c * trans(t)
        for (i = 0; i < 8; i = i + 1) begin
            for (j = 0; j < 8; j = j + 1) begin
                o[8 * i + j] = 9'h0;
                for (k = 0; k < 8; k = k + 1) begin
                    temp29 = mult13_16(c[8 * i + k], t[8 * j + k]);
                    o[8 * i + j] = o[8 * i + j] + temp29[27:18];
                end

            end
        end
        'wait
        // write out the result
        done <= 1'b1;
        while (!start) 'wait 'wait
        for (i = 0; i < 8; i = i + 1) begin
            for (j = 0; j < 8; j = j + 1) begin
                dout <= o[8 * i + j];
                'wait
            end
        end
    end
endmodule
```

## 9.3   Simulation

The design is now ready to simulate. It's helpful to simulate it as soon as possible, because something may have been lost in translation, and this is almost certainly a cleaner HDL description we have after we optimize for synthesis. That is, it will be easier to debug.

Next we need a test bench. The one shown here is for the Verilog. I have not included the VHDL test bench because it seemed like unnecessary clutter: it is on the diskette.

Instead of using the coefficients of the cosine transform, these test benches use an identity matrix. The reason I chose the identity matrix was that it gives me a simple test of whether the indices, etc. are right: if the algorithm is correctly implemented, I should see the same pattern coming out that went in.

The test pattern used here is not very interesting: its most important property is that it is asymmetric and so can be used to test for transposition, rotation, and mirroring transformations.

```verilog
'include "idct.v"
module testbench;
   reg clk, reset, start;
   reg  [11:0] instream;
   reg  [11:0] sample;
   wire [8:0] outstream;
   wire done;
   integer i, j;
   idct fourier (clk, reset, start, done, instream, outstream);
   initial begin: po_reset
      reset = 1'b1;
      reset = #102 1'b0;
   end
   initial begin: clocking
      clk = 1'b0;
      forever begin
         #50 clk = ~clk;
      end
   end
   initial begin: stimulus
      sample <= 12'hded;
      @(posedge clk);
      @(negedge reset);
      start <= 1'b1;
      for (i = 0; i < 8; i = i + 1) begin
          for (j = 0; j < 8; j = j + 1) begin
              if (i < 2 || i == 7) begin
                  sample = 12'h0;
              end
              else if (i == 2)  begin
                      if (j == 0 || j == 6 || j == 7) begin
                          sample = 12'h0;
                      end else begin
                          sample = 12'h1ff;
                      end
                  end
```

```
                else if (i == 3) begin
                        if (j == 2 || j == 5 || j == 6) begin
                            sample = 12'h1ff;
                        end else begin
                            sample = 12'h0;
                        end
                    end
                else if (i == 4) begin
                        if (j == 2 || j == 5) begin
                            sample = 12'h1ff;
                        end else begin
                            sample = 12'h0;
                        end
                    end
                else if (i == 5) begin
                        if (j < 2 || j > 5) begin
                            sample = 12'h0;
                        end else begin
                            sample = 12'h1ff;
                        end
                    end
                else begin // i = 6
                        if (j == 3 || j ==4) begin
                            sample = 12'h1ff;
                        end else begin
                            sample = 12'h0;
                        end
                    end

            instream <= sample;
            @(posedge clk); #2;
        end
    end
    sample <= 12'hded;
    @(posedge clk);
    @(posedge done);
    start = 1'b1;
    @(posedge clk);
    for (i = 0; i < 8; i = i + 1) begin
        for (j = 0; j < 8; j = j + 1) begin
            @(posedge clk);
            $display("out[%d][%d] = %x at %t ns", i, j, outstream, $time);
        end
    end
```

```
    $stop;
  end
endmodule
```

The HDL texts given above are straightforward translations of the C code into Verilog and VHDL. They are clean, consistent, and good for simulation, but not really suitable for synthesis. The reasons for avoiding this description were discussed in general terms above; let's look at the code in the light of the general remarks and see what we can do to improve it.

First, the memories are going to be a major bottleneck. This wouldn't be the case in C, because C runs on a single processor with one main memory, a cache, a limited number of registers, and is compiled by an optimizing compiler: so accessing memory in each pass of a loop is not something we normally worry about. In BC, by contrast, in each of the matrix multiplications the innermost loop as it is written is unparallelizable, because of the memory writes to the result matrices. These writes set up dependencies that in effect serialize the inner loops. Moreover, we can easily eliminate them by using the temporary variables. Thus this HDL code

```
    o[8 * i + j] = 9'h0;
    for (k = 0; k < 8; k = k + 1) begin
        temp29 = mult13_16(c[8 * i + k], t[8 * j + k]);
        o[8 * i + j] = o[8 * i + j] + temp29[27:18];
    end
```

becomes this HDL code, eliminating the memory write bottleneck.

```
    temp29 = 29'b0;
    for (k = 0; k < 8; k = k + 1) begin
        temp29 = temp29 +
                mult13_16(c[8 * i + k], t[8 * j + k]);
    end
    o[8 * i + j] = temp29[27:18];
```

The second form is much better than the first form, because the first has write-write dependencies between the accesses of the memory o. By using a temporary variable instead, we have made it possible to perform the memory write once for each element of o, instead of eight times. Rewriting in this way, if it can be done, is generally superior to removing the write-write dependencies by use of the ignore_memory_precedences (*q. v.*) command, because it reduces the memory bandwidth requirement while ignore_memory_precedences does not.

The other memory accesses of the inner loop are reads: no dependencies exist between them. Thus they can be performed in parallel. This can be done in two ways: first, we can use a multiport memory; or second, we can use a memory with a larger word width and pack two or more data elements in a word. Again, in C this was not something we needed to think about; but we have more freedom now.

Multiport memories are often useful because random accesses can be performed simultaneously without any constraints on the addresses being accessed. Suppose, for example, we had a computation that included these statements somewhere.

```
x = mem[i];
y = mem[j];
```

Two reads are never in conflict, so there is no reason BC could not schedule these two reads independently of one another; and as a special case, they could be scheduled in the same cstep. If the memory has two ports, then the accesses can proceed in parallel, regardless of the values of i and j. By contrast, a wide-word memory gives us increased bandwidth only if we can use the data as it is packed.

So there are two ways to relax the bottleneck in our matrix multiplications: we can use an $n$-port memory or we can use a memory whose word size is $n$ times the size of our data words. Depending on the cost parameters, desired bandwidth, availability of different kinds of memories, and the access patterns of the rest of the design, we might go either way on this. Let's look at wider word sizes.

The bundling requirement means that it is necessary to reorganize the memory to use wide-word access, so that we can get multiple data samples in one word. Suppose we want to get two samples per word; here is the second matrix multiplication organized in that way. The first multiplication is a little more complicated, so I'll delay that until we've gotten the simple one out of the way.

Here is the VHDL for the improved version of the second multiplication.

```
type ct is array (31 downto 0) of signed (25 downto 0);
type tt is array (31 downto 0) of signed (31 downto 0);
...
variable temp26: signed (25 downto 0);
variable temp29: signed (28 downto 0);
variable temp32: signed (31 downto 0);
...
for i in 0 to 7 loop
   for j in 0 to 7 loop
      temp29 := (others => '0');
      for k in 0 to 3 loop
         temp32 := t(4 * j + k);
         temp26 := c(4 * i + k);
         temp29 := temp29 +
               temp26(25 downto 13) * temp32(31 downto 16) +
               temp26(12 downto 0) * temp32(15 downto 0);
      end loop;
      o(8 * i + j) := temp29(26 downto 18);
   end loop;
end loop;
```

To do this I had to redefine the types of the memory variables and create some
new temporaries. The temporaries may or may not end up mapped to registers: it
depends on whether they need to be stored or can be chained.

Here is the Verilog for the second multiplication of the improved version.

```
reg [25:0] temp26;
reg [28:0] temp29;
reg [31:0] temp32;
reg [15:0] t [31:0];
reg [25:0] c [31:0];
...
for (i = 0; i < 8; i = i + 1) begin
   for (j = 0; j < 8; j = j + 1) begin
       temp29 = 29'b0;
       for (k = 0; k < 4; k = k + 1) begin
           temp32 = t[4 * j + k];
           temp26 = c[4 * i + k];
           temp29 = temp29 +
                   mult13_16(temp26[25:13], temp32[31:16]) +
                   mult13_16(temp26[12:0], temp32[15:0]);
       end
       o[8 * i + j] = temp29[27:18];
   end
end
```

Now the memory width has been doubled, that each member of the coefficient
memory actually contains two coefficients. The temporary memory has also had its
bitwidth doubled, so that it contains two adjacent members of the same row in each
word. It is organized that way because the rows are accessed in the transpose-and-
multiply operation. This means that we can perform two multiplications for each
access to the coefficient matrix.

We have to make the code a little more complicated than that for the first matrix
multiply; the problem is that $t_{ij}$ and $t_{ij+1}$ are stored in the same word, and we don't
want to do a read-modify-write cycle on the temporary matrix. Thus we should
compute the elements $t_{ij}$ and $t_{ij+1}$ in the same pass through the inner loop. This
in turn means we have to access the $j^{th}$ and $j + 1^{st}$ rows of the input data in order
to compute the $j^{th}$ and $j + 1^{st}$ columns of the $i^{th}$ row of the temporary matrix t.

We also have to change the organization of the input array x so that the first
multiplication doesn't have to do two memory accesses. Again, I'll use bigger words
with two samples apiece. We will also need a few new temporary variables; BC
does variable lifetime analysis and register sharing, so the new temporary variables
need not add any storage to the final design.

Here is the first loop as reorganized in VHDL.

```vhdl
type xt is array (31 downto 0) of signed (23 downto 0) ;
...
variable temp24: signed (23 downto 0);
variable temp25a: signed (24 downto 0);
variable temp25b: signed (24 downto 0);
variable temp26: signed (25 downto 0);
variable x: xt;
...
for i in 0 to 7 loop
   for j in 0 to 3 loop
       temp25a := (others => '0');
       temp25b := (others => '0');
       for k in 0 to 3 loop
           temp26 := c(4 * i + k);
           temp24 := x(4 * (j * 2) + k);
           temp25a := temp25a +
               temp26(25 downto 13) * temp24(23 downto 12) +
               temp26(12 downto 0) * temp24(11 downto 0);
           temp24 := x(4 * (j * 2 + 1) + k);
           temp25b := temp25b +
               temp26(25 downto 13) * temp24(23 downto 12) +
               temp26(12 downto 0) * temp24(11 downto 0);
       end loop;
       temp32(31 downto 16) := temp25a(23 downto 8);
       temp32(15 downto 0) := temp25b(23 downto 8);
       t(4 * i + j) := temp32;
   end loop;
end loop;
```

This is the Verilog for the new first multiplication.

```verilog
reg [23:0] temp24;
reg [24:0] temp25a;
reg [24:0] temp25b;
reg [25:0] temp26;
reg [23:0] x [31:0];
...
for (i = 0; i < 8; i = i + 1) begin
   for (j = 0; j < 4; j = j + 1) begin
       temp25a = 25'b0;
       temp25b = 25'b0;
       for (k = 0; k < 4; k = k + 1) begin
           temp26 = c[4 * i + k];
```

```
            temp24 = x[4 * (j * 2) + k];
            temp25a = temp25a +
                    mult13_12(temp26[25:13], temp24[23:12]) +
                    mult13_12(temp26[12:0], temp24[11:0]);
            temp24 = x[4 * (j * 2 + 1) + k];
            temp25b = temp25b +
                    mult13_12(temp26[25:13], temp24[23:12]) +
                    mult13_12(temp26[12:0], temp24[11:0]);
        end
        t[4 * i + j] = {temp25a[23:8], temp25b[23:8]};
    end
end
```

Now BC can use single-port memories to construct a schedule with 128 passes through the inner loop of the first multiplication and 256 passes through the inner loop of the second. Notice that we have only *potentially* improved the speed of the design: it could still be scheduled with many csteps per pass. In fact, the minimum-latency schedule of the first multiplication's inner loop is longer than it was, because it is necessary to do two memory accesses now. However, the increase in parallelism in the CDFG now makes it possible to schedule the design with roughly the same number of csteps (static latency) and fewer passes through the innermost loop (dynamic latency): that's where we've gained.

This kind of optimization can go on until we have a single pass for each and all 512 multiplications happening in parallel. Exactly where the tradeoff is best depends on the design requirements. Of course in performing the tradeoff analysis it is necessary also to take the memory latency into account; the tradeoff is different if you have a two-cycle memory access instead of a single-cycle access.

Another point that is not always immediately apparent is my use of indices that can be multiplied by powers of two. BC is fairly clever about optimizing multiplications, but you still aren't going to get a much cheaper multiplication than a left shift. Thus the two-dimensional array accesses such as x[4 * (j * 2 + 1) + k] are actually not as expensive as they look, even if the loops remain rolled. If the loops are unrolled, then the addressing computations turn into constant expressions and can be optimized out altogether.

So let's consider unrolling the loops. Recall that BC will by default unroll all **for** loops with fixed iteration bounds. Hence the second loop will still take about 256 cycles (assuming a single-cycle memory access) because of the memory bottleneck; but it will also take a long time to schedule, because BC will be dealing with 512 multiplications, 512 additions, and 320 memory accesses for each matrix multiplication. That is too large a problem for BC to handle efficiently; the recommendation is that no single scheduling problem (i.e. rolled loop) should contain more than 200 or so schedulable operations, and BC will run faster if you have fewer operations. You might or might not save hardware by unrolling: you will be adding a lot of states to the control FSM, and this will offset the savings in the addressing and loop counting hardware (which are only three- and four-bit operations anyway).

Furthermore, because the memories are the bottleneck, nothing can make the design go faster than the memories can supply and store data. Thus there is not much to be gained in allowing the outer loops (on i and j) to be unrolled; and for the innermost loops there is no certain benefit either way. Hence we will leave the inner loop of each matrix multiplication unrolled but put the **dont_unroll** flag on the outer ones. We will also keep the loops that read the input and write the output rolled; unrolling them would increase scheduling run time with no clear benefit[1].

For synthesis, it is also necessary in the Verilog to move the declarations of the memories down into the process; BC doesn't support global memories in Verilog unless they are mapped to registers. This in turn necessitated moving the initialization of the coefficient ROM into the process block; again, I used the **translate_off** pragma to prevent this part of the design's behavior from being synthesized.

I also had to add a pair of clock edge statements in order to make the design legal for superstate-mode scheduling. These two clock edges separate the writes to the **done** strobe from their immediate successors, which are handshaking loops that wait for the **start** signal to go true. Without these clock edge statements, the design was in violation of the superstate scheduling rule that requires a clock edge separating a write and a loop whose first superstate contains any I/O.

The last thing I did to make this description synthesizable was to add memory mapping pragmas so that the memories would be implemented using RAMs. This is necessary because in the absence of the pragmas the storage arrays would have turned into huge registers, with some potentially ugly multiplexing and slicing logic wrapped around them. This would not only cause very long run times, but it would also result in markedly inferior logic.

The memories I used were of my own construction; their names are of the form **my**$ww$**x**$bb$**r**$y$**m**, where the substrings $ww$ and $bb$ give the number of words and bit width of a word respectively, and the character $y$ is either **o**, signifying a ROM, or **a**, signifying a RAM. The reason I used homebrew memories was that I wanted to give you a concrete example of how RAM wrappers are constructed and used; the complete example is included in Appendix A. The RAMs I have used here all have two-cycle pipelined read capability, as does the ROM: the ROM is pessimistic in that respect.

Here is the synthesizable VHDL for the optimized design.

```
library ieee;
use ieee.std_logic_1164.all;
use ieee.std_logic_arith.all;

package idct_types is
    type xt is array (31 downto 0) of signed (23 downto 0) ;
```

---

[1] As a matter of fact, there might be a reason to partially unroll the I/O loops in the source code. We could then interleave the data input with the head of the first matrix multiplication and the data output with the tail of the second. Such a strategy would unfortunately lead to rather obtuse-looking code. Try it!

```vhdl
   type ct is array (31 downto 0) of signed (25 downto 0) ;
   type ot is array (63 downto 0) of signed (8 downto 0) ;
   type tt is array (31 downto 0) of signed (31 downto 0) ;
end idct_types;

library ieee;
use ieee.std_logic_1164.all;
use ieee.std_logic_arith.all;
library synopsys;
use synopsys.attributes.all;
use work.idct_types.all;

entity idct is
 port (clk, reset, start: in std_logic;
       done: out std_logic;
       din: in signed (11 downto 0);
       dout: out signed (8 downto 0));
end idct;

architecture beh of idct is
begin
   main: process
      variable temp12: signed (11 downto 0);
      variable temp13a: signed (12 downto 0);
      variable temp13b: signed (12 downto 0);
      variable temp24: signed (23 downto 0);
      variable temp25a: signed (24 downto 0);
      variable temp25b: signed (24 downto 0);
      variable temp26: signed (25 downto 0);
      variable temp29: signed (28 downto 0);
      variable temp32: signed (31 downto 0);
      variable x: xt;
      variable c: ct;
      variable o: ot;
      variable t: tt;

      subtype resource is integer;
      attribute variables : string;
      attribute map_to_module : string;

      -- Declare a resource to attribute the RAM
      constant xram : resource := 0;
      attribute variables of xram : constant is "x";
      attribute map_to_module of xram : constant is "my32x24ram";
```

```vhdl
    constant cram : resource := 0;
    attribute variables of cram : constant is "c";
    attribute map_to_module of cram : constant is "my32x26rom";
    constant oram : resource := 0;
    attribute variables of oram : constant is "o";
    attribute map_to_module of oram : constant is "my64x9ram";
    constant tram : resource := 0;
    attribute variables of tram : constant is "t";
    attribute map_to_module of tram : constant is "my32x32ram";

    attribute dont_unroll: boolean;
    attribute dont_unroll of o1: label is TRUE;
    attribute dont_unroll of i1: label is TRUE;
    attribute dont_unroll of o2: label is TRUE;
    attribute dont_unroll of m2: label is TRUE;
    attribute dont_unroll of o3: label is TRUE;
    attribute dont_unroll of m3: label is TRUE;
    attribute dont_unroll of o4: label is TRUE;
    attribute dont_unroll of i4: label is TRUE;
begin
  reset_loop: loop

    -- for simulation purposes I have to load up the ROM
    -- I will use an identity matrix just for testing
    -- synopsys translate_off
    load_rom: for i in 0 to 7 loop
       for j in 0 to 3 loop
          if (i = j * 2) then
             temp13a := ('0', '1', others=> '0'); -- 0800;
          else
             temp13a := (others=> '0');
          end if;
          if (i = j * 2 + 1) then
             temp13b := ('0', '1', others=> '0'); -- 0800;
          else
             temp13b := (others=> '0');
          end if;
          temp26(25 downto 13) := temp13a;
          temp26(12 downto 0) := temp13b;
          c(4 * i + j) := temp26;
       end loop;
    end loop;
    -- synopsys translate_on
```

```
done <= '0';
wait until clk'event and clk = '1';
if reset = '1' then exit reset_loop; end if;
while (start /= '1') loop
    wait until clk'event and clk = '1';
    if reset = '1' then exit reset_loop; end if;
end loop;
wait until clk'event and clk = '1';
if reset = '1' then exit reset_loop; end if;

-- Load the Input RAM
o1 : for i in 0 to 7 loop
    i1: for j in 0 to 3 loop
        temp12 := din;
        wait until clk'event and clk = '1';
        if reset = '1' then exit reset_loop; end if;
        temp24(23 downto 12) := temp12;
        temp24(11 downto 0) := din;
        x(i * 4 + j) := temp24;
        wait until clk'event and clk = '1';
        if reset = '1' then exit reset_loop; end if;
    end loop;
end loop;
wait until clk'event and clk = '1';
if reset = '1' then exit reset_loop; end if;
-- first multiplication. t = c * trans(x)
o2 : for i in 0 to 7 loop
    m2: for j in 0 to 3 loop
        temp25a := (others => '0');
        temp25b := (others => '0');
        i2: for k in 0 to 3 loop
            temp26 := c(4 * i + k);
            temp24 := x(4 * (j * 2) + k);
            temp25a := temp25a +
             temp26(25 downto 13) * temp24(23 downto 12) +
             temp26(12 downto 0) * temp24(11 downto 0);
            temp24 := x(4 * (j * 2 + 1) + k);
            temp25b := temp25b +
             temp26(25 downto 13) * temp24(23 downto 12) +
             temp26(12 downto 0) * temp24(11 downto 0);
        end loop;
        temp32(31 downto 16) := temp25a(23 downto 8);
        temp32(15 downto 0) := temp25b(23 downto 8);
        t(4 * i + j) := temp32;
```

```
            end loop;
          end loop;
          wait until clk'event and clk = '1';
          if reset = '1' then exit reset_loop; end if;
          -- second multiplication. o = c * trans(t)
          o3 : for i in 0 to 7 loop
             m3: for j in 0 to 7 loop
                  temp29 := (others => '0');
                  i3: for k in 0 to 3 loop
                      temp32 := t(4 * j + k);
                      temp26 := c(4 * i + k);
                      temp29 := temp29 +
                        temp26(25 downto 13) * temp32(31 downto 16) +
                        temp26(12 downto 0) * temp32(15 downto 0);
                  end loop;
                  o(8 * i + j) := temp29(26 downto 18);
             end loop;
          end loop;
          wait until clk'event and clk = '1';
          if reset = '1' then exit reset_loop; end if;
          -- write out the result
          done <= '1';
          wait until clk'event and clk = '1';
          if reset = '1' then exit reset_loop; end if;
          while (start /= '1') loop
             wait until clk'event and clk = '1';
             if reset = '1' then exit reset_loop; end if;
          end loop;
          wait until clk'event and clk = '1';
          if reset = '1' then exit reset_loop; end if;
          o4 : for i in 0 to 7 loop
             i4: for j in 0 to 7 loop
                  dout <= o(8 * i + j);
                  wait until clk'event and clk = '1';
                  if reset = '1' then exit reset_loop; end if;
             end loop;
          end loop;
       end loop;  -- reset
    end process;
end beh;
```

Here is the optimized Verilog for the IDCT.

```
'define wait begin                          \
   @(posedge clk);                          \
   if (reset) disable reset_loop;  \
 end

module idct(clk, reset, start, done, din, dout);
   input clk, reset, start;
   input [11:0] din;
   output done;
   reg     done;
   output [8:0] dout;
   reg     [8:0] dout;

   reg [3:0] i,j,k;
   reg [11:0] temp12;
   reg [12:0] temp13a;
   reg [12:0] temp13b;
   reg [23:0] temp24;
   reg [24:0] temp25a;
   reg [24:0] temp25b;
   reg [25:0] temp26;
   reg [28:0] temp29;
   reg [31:0] temp32;

   function [24:0] mult13_12;
     /* twos complement multiplication */
     input [12:0] A; // the port names A, B, Z
     input [11:0] B; // are taken from the designWare
                     // otherwise this won't work
     reg sgn;
       // synopsys map_to_operator MULT_TC_OP
       // synopsys return_port_name Z
       // the above tells how to map this whole function
       // to one designWare function.
     begin
       sgn = A[12] ^ B[11];
       if (A[12] == 1'b1) A = ~A + 1'b1;
       if (B[11] == 1'b1) B = ~B + 1'b1;
       mult13_12= A * B;
       if (sgn == 1'b1) mult13_12 = ~mult13_12 + 1'b1;
     end
   endfunction
```

```
function [24:0] mult13_16;
  /* twos complement multiplication */
  input [12:0] A; // the port names A, B, Z
  input [15:0] B; // are taken from the designWare
                  // otherwise this won't work
  reg sgn;
    // synopsys map_to_operator MULT_TC_OP
    // synopsys return_port_name Z
    // the above tells how to map this whole function
    // to one designWare function.
  begin
    sgn = A[12] ^ B[15];
    if (A[12] == 1'b1) A = ~A + 1'b1;
    if (B[15] == 1'b1) B = ~B + 1'b1;
    mult13_16 = A * B;
    if (sgn == 1'b1) mult13_16 = ~mult13_16 + 1'b1;
  end
endfunction

always begin: reset_loop
  reg [23:0] x [31:0];
  reg [25:0] c [31:0];
  reg [8:0]  o [63:0];
  reg [31:0] t [31:0];
/* synopsys resource xram: variables = "x",
                           map_to_module = "my32x24ram"; */
/* synopsys resource crom: variables = "c",
                           map_to_module = "my32x26rom"; */
/* synopsys resource oram: variables = "o",
                           map_to_module = "my64x9ram"; */
/* synopsys resource tram: variables = "t",
                           map_to_module = "my32x32ram"; */


// for simulation purposes I have to load up the ROM
// I will use an identity matrix just for testing
/* synopsys translate_off */
  for (i = 0; i < 8; i = i + 1) begin
    for (j = 0; j < 4; j = j + 1) begin
      if (i == j * 2) temp13a = 13'h0800;
      else temp13a = 13'h0000;
      if (i == j * 2 + 1) temp13b = 13'h0800;
      else temp13b = 13'h0000;
      c[4 * i + j] = {temp13a, temp13b};
```

```
          end
   end
/* synopsys translate_on */

   done <= 1'b0;
   'wait
   while (!start) 'wait 'wait // loop wait plus tail wait
   // Load the Input RAM
   for (i = 0; i < 8; i = i + 1) begin: o1
      /* synopsys resource r1: dont_unroll = "o1"; */
      for (j = 0; j < 4; j = j + 1) begin: i1
         /* synopsys resource r1: dont_unroll = "i1"; */
         temp12 = din;
         'wait
         x[i * 4 + j] = {temp12, din};
         'wait
      end
   end
   'wait
   // first multiplication. temp = c * trans(x)
   for (i = 0; i < 8; i = i + 1) begin: o2
      /* synopsys resource r1: dont_unroll = "o2"; */
      for (j = 0; j < 4; j = j + 1) begin: m2
         /* synopsys resource r1: dont_unroll = "m2"; */
         temp25a = 25'b0;
         temp25b = 25'b0;
         for (k = 0; k < 4; k = k + 1) begin: i2
             temp26 = c[4 * i + k];
             temp24 = x[4 * (j * 2) + k];
             temp25a = temp25a +
                   mult13_12(temp26[25:13], temp24[23:12]) +
                   mult13_12(temp26[12:0], temp24[11:0]);
             temp24 = x[4 * (j * 2 + 1) + k];
             temp25b = temp25b +
                   mult13_12(temp26[25:13], temp24[23:12]) +
                   mult13_12(temp26[12:0], temp24[11:0]);
         end
         t[4 * i + j] = {temp25a[23:8], temp25b[23:8]};
      end
   end
   'wait
   // second multiplication. o = c * trans(t)
   for (i = 0; i < 8; i = i + 1) begin: o3
      /* synopsys resource r1: dont_unroll = "o3"; */
```

```
            for (j = 0; j < 8; j = j + 1) begin: m3
                /* synopsys resource r1: dont_unroll = "m3"; */
                temp29 = 29'b0;
                for (k = 0; k < 4; k = k + 1) begin: i3
                    temp32 = t[4 * j + k];
                    temp26 = c[4 * i + k];
                    temp29 = temp29 +
                            mult13_16(temp26[25:13], temp32[31:16]) +
                            mult13_16(temp26[12:0], temp32[15:0]);
                end
                o[8 * i + j] = temp29[26:18];
            end
        end
        'wait
        // write out the result
        done <= 1'b1;
        'wait
        while (!start) 'wait 'wait
        for (i = 0; i < 8; i = i + 1) begin: o4
            /* synopsys resource r1: dont_unroll = "o4"; */
            for (j = 0; j < 8; j = j + 1) begin: i4
                /* synopsys resource r1: dont_unroll = "i4"; */
                dout <= o[8 * i + j];
                'wait
            end
        end
    end
endmodule
```

The synthesis scripts I used are amalgamated into this:

```
synthetic_library = {dw01.sldb dw02.sldb dw03.sldb myrams.sldb}
search_path = search_path + /u/dknapp/dw/src
/* this search path needs to be modified */
/* see appendix A for directions on building your own
   DesignWare directory. */
analyze -f verilog step4.v
/* analyze -f vhdl step4.vhd */
elaborate -s idct
create_clock clk -p 50
schedule -io su
report_schedule -su > idct_sum
report_schedule -ab > idct_fsm
report_schedule -op > idct_ops
```

The summary report shows the following loop structures and area summary:

```
Loop timing information:
    reset_loop........................28 cycles (cycles 0 - 28)
        loop_101......................1 cycle  (cycles 1 - 2)
            (exit) EXIT_L101...................  (cycle 2)
        o1............................3 cycles (cycles 3 - 6)
            (exit) EXIT_L103...................  (cycle 4)
        i1............................2 cycles (cycles 4 - 6)
            (exit) EXIT_L105...............  (cycle 5)
        o2............................9 cycles (cycles 6 - 15)
            (exit) EXIT_L115...................  (cycle 7)
        m2............................8 cycles (cycles 7 - 15)
            (exit) EXIT_L117...............  (cycle 8)
        o3............................6 cycles (cycles 15 - 21)
            (exit) EXIT_L137...................  (cycle 16)
        m3............................5 cycles (cycles 16 - 21)
            (exit) EXIT_L139...............  (cycle 17)
        loop_156......................1 cycle  (cycles 23 - 24)
            (exit) EXIT_L156...................  (cycle 24)
        o4............................3 cycles (cycles 25 - 28)
            (exit) EXIT_L157...................  (cycle 26)
        i4............................2 cycles (cycles 26 - 28)
            (exit) EXIT_L159...............  (cycle 27)
--------------------------------------------------------------
    Area Summary
--------------------------------------------------------------
Estimated combinational area    8066
Estimated sequential area       1470
TOTAL                           9536
```

The important things to notice here are that the two multiplication loops (o2 and o3) are taking up latencies of eight and five cycles respectively for their inner loops; these inner loops, recall, will be executed 128 and 256 times respectively. Thus a rough guess at the throughput, exclusive of loading and unloading the input and output matrices, leads us to a performance of roughly $32 \times 8 + 64 \times 5 = 576$ cycles per input matrix. This is considerably better than what I would have gotten from the original HDL: my guess is about 2000 after I took out the memory write dependency. Part of the gain in performance was my reorganization of the memory, which roughly doubled the potential throughput. The other part was the fact that the innermost loop of each multiplication was unrolled, which in conjunction with pipelined memory read cycles allowed BC to cram much more into the inner loops than would otherwise have been possible.

Here is part of the state diagram; I have taken out all but the two matrix multiplications so that you can see the actual number of state transitions in each of the multiplication loops. The first matrix multiplication is represented by the states s_4_7 through s_5_15; the second multiplication is represented by s_6_16 through s_7_21. I have abstracted the operations in order to save space. The full reports are included on the diskette.

| | | | |
|---|---|---|---|
| s_4_7 | 0-- | s_5_8 | (mem, mul, add, compare) |
| s_4_7 | 1-- | s_6_16 | (add, compare indices) |
| s_5_8 | 01- | s_4_7 | (add, compare indices) |
| s_5_8 | 00- | s_5_9 | (many operations) |
| s_5_9 | 00- | s_5_10 | (many operations) |
| s_5_10 | 00- | s_5_11 | (many operations) |
| s_5_11 | 00- | s_5_12 | (many operations) |
| s_5_12 | 00- | s_5_13 | (many operations) |
| s_5_13 | 00- | s_5_14 | (many operations) |
| s_5_14 | 00- | s_5_15 | (many operations) |
| s_5_15 | 00- | s_5_8 | (many operations) |
| . | . | . | |
| s_6_16 | 0-- | s_7_17 | (mem, mul, add, compare) |
| s_7_17 | 0-1 | s_6_16 | (add, compare indices) |
| s_7_17 | 0-0 | s_7_18 | (many operations) |
| s_7_18 | 0-0 | s_7_19 | (many operations) |
| s_7_19 | 0-0 | s_7_20 | (many operations) |
| s_7_20 | 0-0 | s_7_21 | (many operations) |
| s_7_21 | 0-0 | s_7_17 | (many operations) |

To improve the performance even further, I would go on to reorganize the memories into four-word records or eight-word records. I could also build a new combinational or pipelined DesignWare part (see Appendix A) that would perform a carry-save sum-of-products multiply and accumulate. Such a part would increase the throughput of the design even more, if I had the memory bandwidth to support it. Pipelining the new part would allow me to decrease the clock rate until I hit the minimum memory clock cycle time.

### Summary

In this chapter we have translated a highly abstract mathematical specification first into C and then into VHDL and Verilog. The initial translation into HDL was straightforward but somewhat naive: we then proceeded to improve it. The chapter emphasizes the reasoning behind each transformation of the naive implementation, and the differences between C and hardware that make the transformations compelling. Finally, we synthesized the design using DesignWare RAMs the construction of which are described in Appendix A.

# Chapter 10

# The Data Encryption Standard: Random Logic

In this chapter the design problem is a circuit that implements the Data Encryption Standard (DES) [17], [3], [19]. The circuit will encrypt and decrypt eight-byte character strings, using another eight-byte string as a key. This circuit demonstrates a design with almost no arithmetic in the datapath, not very much control logic, and a lot of low-level bit manipulation.

The most interesting thing about the DES from our point of view is that we can still use behavioral synthesis to construct an efficient implementation. This design is a counterexample to a common misconception about behavioral synthesis: that it is only well suited to algorithms with a lot of arithmetic and shared hardware.

## 10.1  General Description

The DES consists of a series of permutations and lookup table accesses, and is almost entirely built out of random logic. It takes in a 64-bit key, and a 64-bit data vector, then performs the permutations and table lookups in a loop with a variable number of passes. Here the number of passes is 16; it is controlled by the constant parameter ROUNDS.

In each round the function **f**, which is the DES core function, is called once. The function **f** in turn calls a function **p**, which constructs a permutation of the results of eight calls to a function **sbox**. The **sbox**, which is just a lookup table, is therefore called a total of 128 times, and is a good candidate for sharing.

Sharing random logic requires the user to take a number of specific actions. The logic must be encapsulated as a function, which is then mapped to a Design-Ware module; pragmas must be used to direct BC to use the DesignWare module instead of inlining the function call. Because the module has by this process become synthetic, it can be shared.

Sharing isn't always a good thing, though. The **sbox** is a very simple function. It consists of a case statement, which chooses a lookup table. The table is chosen on the basis of one of the **sbox** function's input parameters; this parameter is always a constant. The `elaborate` command does constant propagation: so when the inlined function is elaborated, the unused case clauses and their tables will evaporate.

Making the `sbox` sharable, instead of inlined, means that the case statement and lookup table logic will be preserved in a single complex unit. Thus the sharable synthetic unit will be much more complex than any of the inlined units; unless it can indeed be shared there will be no savings except from the partitioning induced by the hierarchy of synthetic modules. And for each case where the function is shared, there will be additional muxing costs; so we should expect to gain by encapsulation in cases where there is a lot of sharing, i.e. in designs with high latency. Notice, however, that these considerations apply to the `sbox` function: if we were thinking about a more complex function we might make other choices.

Preserving the `sbox` function and constructing a DesignWare version will have the same effect, because the function contains no 'behavioral' constructs.

In the HDL text as it was originally written, the loop `for1`, which performs one of the sixteen rounds on each pass, would be unrolled by default. Inside each pass of `for1` is a call to the DES function `f`; and each call of `f` has eight calls to the random-logic function `sbox`. This design would therefore have 128 copies of the inlined sbox function. That's a lot of logic, leading to a high-cost implementation and/or a long run time.

There are four things that can be done about this. First, you can use a `dont_unroll` on the main loop of the algorithm. That's present in the code below and on the diskette; it cuts the number of operations being scheduled by a factor of about sixteen.

Second, try building a DesignWare or preserved version of `sbox`.

Third, build a DesignWare or preserved version of `f`.

Finally, try unrolling the `for1` loop either wholly or in part. Unrolling the loop will help to reduce the latency of the design: as it stands, there are 48 cycles total (three cycles per round, 16 rounds). Unrolling the loop with fewer cycles per round may reduce the overall latency and lead to a higher performance. If either `f` or the `sbox` is encapsulated before unrolling, the unrolled loop will be much more tractable than otherwise.

Unrolling the loop completely makes DesignWare encapsulation essential, because the amount and complexity of random logic would otherwise swamp the scheduler. Partial unrolling can be done in any of several ways: for example by constructing two loops of four iterations each, and allowing the inner loop to be unrolled. It isn't clear *a priori* what degree of unrolling is going to help the most; there could very well be an optimum somewhere between completely unrolling the loop and leaving it completely rolled.

## 10.2  HDL Description

On the next page is the Verilog source for the DES. I have included both Verilog and VHDL on the accompanying diskette.

```verilog
'define wait                                               \
    begin                                                 \
      @(posedge clk);                                     \
      if (reset == 1'b1) disable reset_loop; \
    end
   // defines the s1 - s7 constants for the sboxes
'include "smax.v"
   // a lookup table selecting one or two rotations
'define two_rot 16'b0011111101111110
   // number of rounds DES = 16 rounds
'define ROUNDS 5'h10
   // this is the DES module itself
module des(clk, reset, start, done,
           data_in, key_in, decode, data_out);
  input start;
  input [0:63] data_in;
  input [0:63] key_in; // only 56 bits of the key are used
  input decode, clk, reset;
  output [0:63] data_out;
  output done;

  parameter loop_size = 7;
  wire start, decode, clk, reset;
  wire [0:63] data_in;
  wire [0:63] key_in;
  reg [0:63] data_out;
  reg done;

  // des initial permutation of 64 bits
  function [0:63] ip;
    // synopsys return_port_name ip
    input [0:63] i;
    begin
    ip = { i[57],i[49],i[41],i[33],i[25],i[17],i[9], i[1],
           i[59],i[51],i[43],i[35],i[27],i[19],i[11],i[3],
           i[61],i[53],i[45],i[37],i[29],i[21],i[13],i[5],
           i[63],i[55],i[47],i[39],i[31],i[23],i[15],i[7],
           i[56],i[48],i[40],i[32],i[24],i[16],i[8], i[0],
           i[58],i[50],i[42],i[34],i[26],i[18],i[10],i[2],
           i[60],i[52],i[44],i[36],i[28],i[20],i[12],i[4],
           i[62],i[54],i[46],i[38],i[30],i[22],i[14],i[6]  };
    end
  endfunction
```

```
// DES inverse initial permutation
function [0:63] iip;
  // synopsys return_port_name iip
  input [0:63] i;
  begin
  iip = { i[39],i[7], i[47],i[15],i[55],i[23],i[63],i[31],
          i[38],i[6], i[46],i[14],i[54],i[22],i[62],i[30],
          i[37],i[5], i[45],i[13],i[53],i[21],i[61],i[29],
          i[36],i[4], i[44],i[12],i[52],i[20],i[60],i[28],
          i[35],i[3], i[43],i[11],i[51],i[19],i[59],i[27],
          i[34],i[2], i[42],i[10],i[50],i[18],i[58],i[26],
          i[33],i[1], i[41],i[9], i[49],i[17],i[57],i[25],
          i[32],i[0], i[40],i[8], i[48],i[16],i[56],i[24]  };
  end
endfunction

// DES permutation function -- permutes 32 bits
function [0:31] p;
  // synopsys return_port_name p
  input [0:31] i;
  begin
   p = { i[15],i[6], i[19],i[20],i[28],i[11],i[27],i[16],
         i[0], i[14],i[22],i[25],i[4], i[17],i[30],i[9],
         i[1], i[7], i[23],i[13],i[31],i[26],i[2], i[8],
         i[18],i[12],i[29],i[5], i[21],i[10],i[3], i[24] };
  end
endfunction

// DES expansion function -- expands 32 bits to 48 bits
function [0:47] e;
  // synopsys return_port_name e
  input [0:31] i;
  begin
  e = { i[31],i[0], i[1], i[2], i[3], i[4], i[3], i[4],
        i[5], i[6], i[7], i[8], i[7], i[8], i[9], i[10],
        i[11],i[12],i[11],i[12],i[13],i[14],i[15],i[16],
        i[15],i[16],i[17],i[18],i[19],i[20],i[19],i[20],
        i[21],i[22],i[23],i[24],i[23],i[24],i[25],i[26],
        i[27],i[28],i[27],i[28],i[29],i[30],i[31],i[0] };
  end
endfunction
```

```verilog
// permuted choice #1
function [0:55] pc1;
  // synopsys return_port_name pc1
  input [0:63] i;
  begin
    pc1 = { i[56],i[48],i[40],i[32],i[24],i[16],i[8],
            i[0], i[57],i[49],i[41],i[33],i[25],i[17],
            i[9], i[1], i[58],i[50],i[42],i[34],i[26],
            i[18],i[10],i[2], i[59],i[51],i[43],i[35],
            i[62],i[54],i[46],i[38],i[30],i[22],i[14],
            i[6], i[61],i[53],i[45],i[37],i[29],i[21],
            i[13],i[5], i[60],i[52],i[44],i[36],i[28],
            i[20],i[12],i[4], i[27],i[19],i[11],i[3] };
  end
endfunction

// permuted choice #2
function [0:47] pc2;
  // synopsys return_port_name pc2
  input [0:55] i;
  begin
    pc2 = { i[13],i[16],i[10],i[23],i[0], i[4],
            i[2], i[27],i[14],i[5], i[20],i[9],
            i[22],i[18],i[11],i[3], i[25],i[7],
            i[15],i[6], i[26],i[19],i[12],i[1],
            i[40],i[51],i[30],i[36],i[46],i[54],
            i[29],i[39],i[50],i[44],i[32],i[47],
            i[43],i[48],i[38],i[55],i[33],i[52],
            i[45],i[41],i[49],i[35],i[28],i[31] };
  end
endfunction

function two_rotations;
  // synopsys return_port_name two_rotations
  input [3:0] i;
  reg [0:15] two_rot_tbl;
  begin
    two_rot_tbl = `two_rot;
    two_rotations = two_rot_tbl[i];
  end
endfunction
```

```verilog
// function descibing the DES s boxes
function [3:0] sbox;
  // synopsys return_port_name s
  input [5:0] i;
  input [2:0] n;
  reg [7:0] index;
  reg [0:255] tbl;
  begin
    index = { i[5], i[0], i[4:1], 2'b00 };
    case (n)
      0: tbl = 's1; // these constants from smax.v
      1: tbl = 's2;
      2: tbl = 's3;
      3: tbl = 's4;
      4: tbl = 's5;
      5: tbl = 's6;
      6: tbl = 's7;
      7: tbl = 's8;
    endcase
    sbox = {tbl[index],tbl[index+1],tbl[index+2],tbl[index+3]};
  end
endfunction

// the DES function
function [0:31] f;
  // synopsys return_port_name f
  input [0:31] i;
  input [0:47] kk;
  reg [0:47] ik;
  begin : des_f
    ik = e(i) ^ kk;
    f = p( {
             sbox(ik[0:5],   3'h0),
             sbox(ik[6:11],  3'h1),
             sbox(ik[12:17], 3'h2),
             sbox(ik[18:23], 3'h3),
             sbox(ik[24:29], 3'h4),
             sbox(ik[30:35], 3'h5),
             sbox(ik[36:41], 3'h6),
             sbox(ik[42:47], 3'h7)
             } );
  end
endfunction
```

```verilog
reg [0:31] l, r, t;
reg [0:27] c, d;
reg [0:47] k;
reg [5:0] x;
reg [4:0] j;

integer loops;
// DES algorithm
always begin : b1
  // synopsys resource r1:dont_unroll = "for1";
  begin : reset_loop
  done <= 1'b0;
  'wait
  forever begin
    while (start == 1'b0) 'wait
    'wait
    done <= 1'b0;
    {l, r} = ip(data_in); // Do inital permutation
    {c, d} = pc1(key_in); // Init key schedule
    'wait
    // Apply f iteratively for ROUNDS rounds
    for (j=0; j<'ROUNDS; j = j + 1) begin : for1
        'wait
        if (decode == 1'b1) // Update key schedule
        begin
          k = pc2({c,d});
          c = {c[27], c[0:26]};
          d = {d[27], d[0:26]};
          if ( two_rotations('ROUNDS-1-j) == 1'b1)
          begin
            c = {c[27], c[0:26]};
            d = {d[27], d[0:26]};
          end
        end else begin
          c = {c[1:27], c[0]};
          d = {d[1:27], d[0]};
          if ( two_rotations(j) == 1'b1)
          begin
            c = {c[1:27], c[0]};
            d = {d[1:27], d[0]};
          end
          k = pc2({c,d});
        end
        t = r;                // Do a round
```

```
            r = 1 ^ f(r, k);
            l = t;
            for(loops = 0; loops < loop_size; loops = loops + 1) 'wait
          end
          'wait
          // Do inverse initial permutation
          data_out <= iip( {r, l} ); // note final un-swap
          done <= 1'b1; // coversion done!
          while (start == 1'b1) begin
             'wait
          end
          'wait
          done <= 1'b0;
          'wait
      end
      end
    end
endmodule
```

Test benches for the design are also given on the diskette; these test benches encode, then decode, a single eight-character string, using another eight-character string as a key.

## 10.3  Synthesis

Here is the summary report after synthesis of the initial design (rolled main loop and no DesignWare). Notice that the estimated area is about 3500 gates. In fact, when I compiled this design to gates, the area ended up considerably higher than that (about 5500); the reason is that in the version of BC that I used, the area estimation functions don't work all that well on random logic, and this design is mostly random logic. The comparator and adder/subtracter shown below are present because the main loop is rolled; thus it needs explicit loop control hardware. That hardware would be eliminated if the main loop were unrolled; then the only hardware reported by the summary report would be the registers.

```
Clock period 50.00
Loop timing information:
     b1.........................11 cycles (cycles 0 - 11)
        loop_195.................10 cycles (cycles 1 - 11)
           loop_196.............1 cycle  (cycles 1 - 2)
              (exit) EXIT_L196............... (cycle 2)
           for1.................3 cycles (cycles 4 - 7)
              (exit) EXIT_L203............... (cycle 5)
           loop_234.............1 cycle  (cycles 8 - 9)
```

```
                (exit) EXIT_L234............... (cycle 9)
        Area Summary

Estimated combinational area      835
Estimated sequential area        2590
TOTAL                            3425


-----------------------------------------------
        Resource types
-----------------------------------------------
        Register Types
=======================================
            1-bit register....................3
            5-bit register....................2
            28-bit register..................4
            32-bit register..................2

        Operator Types
=======================================
            (5_4->1)-bit DW01_cmp2.............1
            (5_5->5)-bit DW01_addsub..........1
```

## 10.4   Use of DesignWare

The construction of the **sbox** DesignWare proper is detailed in Appendix A. Once the sbox part has been installed in the local DesignWare library (see Appendix A), all that remains to be done is insertion of the **map_to_operator** pragma in the **sbox** function definition of the original HDL text. That looks like this in VHDL.

```
function  sbox (i: bit_vector (5 downto 0);
                n: bit_vector (2 downto 0))
    return bit_vector;
  .    .    .


-- function describing the DES s boxes
function  sbox (i: bit_vector (5 downto 0);
                n: bit_vector (2 downto 0))
      return bit_vector is
  -- pragma return_port_name s
  -- pragma map_to_operator sbox_op
    .    .    .
  end sbox;
```

In Verilog, the `map_to_operator` pragma looks like this.

```
// function descibing the DES s boxes
function [3:0] sbox;
  // synopsys return_port_name s
  // synopsys map_to_operator sbox_op
endfunction
```

The synthesis script must now contain pointers to the new synthetic library:

```
define_design_lib sbx -path ~/dw/lib
search_path = search_path + {~/dw/src}
synthetic_library = synthetic_library + sbx.sldb
analyze -f verilog des_dw.v
/* analyze -f vhdl des_dw.vhd */
elaborate -s des
create_clock clk -p 50
schedule
report_schedule -su > dw_sum
report_schedule -op > dw_ops
report_schedule -ab > dw_abs
```

The calls to **sbox** can now be shared. Notice, however, that the loop **for1** has only three cycles, due to the parameter **loop_size**; there isn't going to be a lot of sharing unless we increase the latency. In my own explorations, I tried values of **loop_size** of two, four, eight, and ten; and got varying degrees of sharing of **sbox** components, all the way from eight sboxes (no sharing) down to one sbox being used eight times in each round.

Having constructed several versions of this design with different latencies and amounts of sbox sharing, the conclusion I drew was that sharing sboxes with the loop still rolled is *not* worth it; in every case where the **for1** loop was kept rolled, the area and performance of the version with shared sboxes was worse.

That isn't the case when the loop is unrolled. In those cases, the amount of unrolling determines the value of the approach. It turns out that the version with the loop unrolled is too complicated to lead to a good design unless the **sbox** logic is encapsulated. Thus encapsulation is the way to achieve the highest throughput for this design.

In order to achieve the high-throughput design, it is necessary to change the number of clock edge statments inside the loop. One way to do this is to set **loop_size** to zero; that gives us as many cycles as we have rounds.

Other versions, with varying amounts of unrolling, can be constructed by the following approach. The single loop **for1** can be broken into two or three nested loops, and then the inner loop(s) can be unrolled. Here is a version with three loops; the first inner loop performs the computation and the second inner loop just provides parametric control of the number of clock edges a single pass through the outer loop will take.

```
// synopsys resource r1:dont_unroll = "for1a";
    .   .   .   .
for (j = 0; j < 4; j = j + 1) begin: for1a
   for (i = 0; i < 4; i = i + 1) begin: for1b
      .   .   .
   end
   for (i = 0; i < desired_latency; i = i + 1) `wait
end
```

Notice that the dont_unroll pragma has been applied to the outer loop, but not the inner loops: the result is that after unrolling we will get a single loop that will be traversed four times, and each time we will have four inlined copies of each of the inner loops. We can then control the latency by removing all clock edges from for1b and setting the parameter desired_latency.

Superstate mode can also be used to accomplish much the same goal. The trick would be to take the excess clock edge statements *out* of the source HDL. Recall that superstate mode may add new clock edges, but will never delete any; thus to improve performance we would have to begin by deleting surplus clock edges from the HDL source. We would then constrain the schedule to have the desired latency by using bc_shell commands.

# Chapter 11

# Packet router

In this chapter we will look at a design for a simple packet router. This circuit demonstrates a design that has only a little arithmetic and logic, but a fair amount of memory; the arithmetic and logic it does have are there to support the memory. As such, it is another counterexample to the common misconception about behavioral synthesis, that behavioral synthesis is really 'for' designs with a lot of arithmetic in a datapath. This design has only the arithmetic necessary for addressing and searching its memory; none of that is particularly complex. What we have instead is a moderately complex control flow in a memory-based design.

Designs with complex control flow are good targets for behavioral synthesis because the control flow can be expressed cleanly by using high-level constructs (loops, conditionals, etc.). This is by contrast with RTL description techniques, which tend to have explicit state-next state mappings built in. Such a state-based RTL description is very much like a program built solely around `goto` statements. As the state diagram becomes complex, it becomes increasingly difficult to understand, debug, and modify the HDL description.

Designs with a lot of memory accesses are also well-suited to behavioral synthesis. The reason for this is that the memory accesses are themselves rather complex events: sometimes taking many cycles, and having particular timing protocols. Thus scheduling and binding memory accesses automatically is a big advantage, especially in cases where the particular memory technology to be used is a variable. Imagine, for example, the kind of RTL specification that uses state-based `goto`s in conjunction with many multicycle memory accesses. Now imagine that, having scheduled the memory access protocols onto the state graph of such a description, your boss tells you to rework the design using a different memory with a different access protocol. Using behavioral synthesis, that would be a matter of changing a single pragma in the source HDL, and (perhaps) changing a latency constraint.

The algorithm that the router executes is built around a *translation lookaside buffer* (TLB) table which is implemented as a memory. In this example, the table has 32 entries. Each entry has two fields: an address, which is 24 bits wide; and a segment number, which is five bits wide. A particular segment address is reserved for broadcast; this address is 31, which is the largest segment address anyway. Thus all recipients are assumed to be reachable on one of 31 segments.

When the router receives a packet the packet consists of three fields. These are a segment address, a recipient address, and a one-bit opcode, that tells the router whether to search for or save the recipient/segment mapping. The packet data is handled externally: it would just get in the way here.

Regardless of the opcode, the router's action on receiving a packet is to find the recipient in the TLB. It does this by binary search; in C, this is approximately

```c
index = 0;
for (i = 4; i >= 0; i++) {
    tmp_index = index + 2 ** i;
    tmp = tlb(tmp_index);
    if (tmp.addr == recipient) {
        found = TRUE;
        result = tmp_index;
        break; /* success */
    }else if (tmp.addr < recipient) {
        index = tmp_index;
    }
}
```

By the time this loop has finished, either the recipient has been found or not. The opcode is either 'search' or 'save'; so there are four cases.

If the opcode is 'search' and the recipient has been found, transmit the segment address that was in the table entry we found.

If the opcode is 'search' and the recipient was not found, broadcast the packet to all segments by using the special segment address 31.

If the opcode is 'save' and the recipient was found, update the recipient's segment address field in the TLB. This would be normal procedure for a return address, for example; thus every time a host sent a packet through this router, the host would have its segment address updated automatically.

If the opcode is 'save' and the recipient was not found, insert the recipient's full address and the segment address that came with the packet into the TLB at the proper point. This in turn requires that the entries currently at that location and above be 'bumped' upward in the memory. That makes insertion of a new address the most expensive thing this particular circuit can do, because the search process is $O(log_2 n)$ in complexity.

In a real router I would also add something to prevent large addresses from falling off the end of the table; perhaps a least-recently-used bit or counter field.

Notice that the run time of this algorithm is unpredictable. Thus it is advisable to provide handshaking; the handshaking we provide is similar to that of the IIR filter example of Chapter 8, with the minor variation that the output strobe toggles to indicate the presence of data, instead of going high and then low again.

Here is the VHDL for the packet router; the Verilog and test bench code is provided on the accompanying diskette.

We begin with a package that defines types, bit widths, and so on.

```
library ieee;
use ieee.std_logic_1164.all;
use ieee.std_logic_arith.all;
package routing is
    subtype     tlb_index_t  is integer range 0 to 31;
    subtype     seg_num_t    is integer range 0 to 31;
    subtype     addr_t       is unsigned (23 downto 0);
    type        opcd is (store,find);
    type        rout_req_t is record
                   address : addr_t;
                   segment : seg_num_t;
                   command : opcd;
                end record;
    type        tlb_entry_t is record
                   address : addr_t;
                   segment : seg_num_t;
                end record;
    type        tlb_t  is array (0 to 31) of tlb_entry_t;
    constant    no_address   : addr_t := (others => '1');
    constant    broadcast    : seg_num_t := 31;
end routing;
```

Here is the VHDL for the router proper.

```
library synopsys;
use synopsys.attributes.all;
library dware;
use dware.behavioral.all;
library ieee;
use ieee.std_logic_1164.all;
use ieee.std_logic_arith.all;
use work.routing.all;

entity rt is
    port (clk          : in  std_logic;
          reset        : in  std_logic;
          inp_strb     : in  std_logic;
          inp          : in  rout_req_t;
          busy         : out std_logic;
          outp_strb    : buffer std_logic;
          outp         : out tlb_entry_t);
end rt;
```

```vhdl
architecture behavior of rt is

begin

  translate : process
    variable    req : rout_req_t;
    variable    tlb_index, old_tlb_index : tlb_index_t;
    variable    response, t1, t2 : tlb_entry_t;
    variable    found : std_logic;
    variable    tlb : tlb_t;

    constant    tlb_ram : resource := 0;
    attribute   variables of tlb_ram : constant is "tlb";
    attribute   map_to_module of tlb_ram :
                constant is "DW03_ram1_s_d";
    attribute   dont_unroll : boolean;
    attribute   dont_unroll of init:label is true;

  begin
  reset_loop: loop
      outp_strb <= '0';
      busy  <= '1';
      wait until clk'event and clk = '1';
      if reset = '1' then exit reset_loop; end if;
      t1.address := no_address;
      t1.segment := broadcast;
 init: for k in tlb'range loop
          tlb(k) := t1;
      end loop init;
      wait until clk'event and clk = '1';
      if reset = '1' then exit reset_loop; end if;
      busy  <= '0';
   endless : loop
          inp_hs : while inp_strb = '0' loop
            wait until clk'event and clk = '1';
            if reset = '1' then exit reset_loop; end if;
          end loop inp_hs;
          req := inp;
          busy  <= '1';
          wait until clk'event and clk = '1';
          if reset = '1' then exit reset_loop; end if;
          tlb_index := 0;
          found := '0';
```

```
search: for i in 4 downto 0 loop
        old_tlb_index := tlb_index;
        tlb_index := tlb_index + tlb_index_t(2**i);
        t1 := tlb(tlb_index);
        if t1.address > req.address then
            tlb_index := old_tlb_index;
        elsif t1.address = req.address then
            found := '1';
            exit search;
        end if;
    end loop search;
    case req.command is
        when find =>
            if found = '1' then
              response := t1;
            else
              response.address := req.address;
              response.segment := broadcast;
            end if;
        when store =>
            response.address := req.address;
            response.segment := req.segment;
            if found = '1' then
                tlb(tlb_index) := response;
            else
                if tlb(tlb_index).address < req.address then
                    tlb_index := tlb_index + 1;
                end if;
                t2 := response;
    insert : while tlb_index < 32 loop
                t1 := tlb(tlb_index);
                tlb(tlb_index) := t2;
                if t1.address = no_address then
                    exit insert;
                end if;
                t2 := t1;
                tlb_index := tlb_index + 1;
            end loop insert;
            end if;
    end case;
    wait until clk'event and clk = '1';
    if reset = '1' then exit reset_loop; end if;
    outp <= response;
    outp_strb <= not outp_strb;
```

```
      busy  <= '0';
      wait until clk'event and clk = '1';
      if reset = '1' then exit reset_loop; end if;
    end loop endless;
  end loop reset_loop;
 end process;
end behavior;
```

Synthesis of this circuit is done using the following script. We have to include some new statements: the synthetic libraries have to be explicitly named to get the DW03 memory required by the memory mapping pragma. I have also added wire loading and operating conditions settings that will result in better logic.

I also had to include the **vhdlout_use_packages** command to ensure that the RTL VHDL written out after synthesis would include the right packages. Otherwise we would get some complaints when we analyzed the RTL for simulation.

```
synthetic_library = { dw01.sldb dw02.sldb dw03.sldb }
analyze -f vhdl p.vhd
analyze -f vhdl rt.vhd
/* analyze -f verilog rt.v */
elaborate -s rt
set_operating_conditions wccom
set_wire_load 10x10 -mode top
create_clock -p 120 clk
schedule -io superstate -area
report_schedule -sum > sum_report
report_schedule -op > rwol_report
report_schedule -op -mask rwL > rwl_report
report_schedule -var > var_report
report_schedule -abs > fsm_report
vhdlout_use_packages = { "IEEE.std_logic_1164",
                         "IEEE.std_logic_arith",
                         "work.routing.all" }
write -h -f vhdl -out pc.vhd
```

Figure 11.1 shows the state diagram (**-abs** report) for this design. Note that the state diagram is fairly complex: imagine writing this schedule by hand, using explicit states and next states. Take a look at the textual version on the diskette: I couldn't in good conscience include it here because it is too long.

Note that the state diagram has nine states that are not shown; that's where the search loop got unrolled. If we kept that loop rolled, the number of states would be decreased by about that amount.

Another benefit of keeping the search loop rolled would be a simplification of the condition vectors. The state machine has 13 inputs and a very complex set of conditions on some of the arcs of the state diagram; the reason is that the loop was

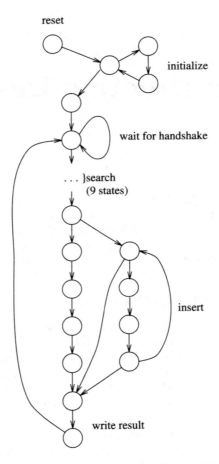

**Figure 11.1.** State diagram of the packet router

unrolled, so the conditionals all became parallel to one another. Thus the number of status bits can potentially explode, as the status bits need to be saved until the last operation that depends on them has been performed. By leaving the loop rolled, this would be reduced because of the serialization implicitly imposed by the loop. The number of states would be reduced as well. In a design like this one, where the control is complex and the datapath is simple, the control FSM might very well come to dominate the overall cost or clock cycle time; thus a reduction in the FSM's input space and its state count might well represent a net saving, in spite of the 'extra' incrementer and comparator needed to drive the loop.

# Appendix A

# Constructing DesignWare

This appendix gives a brief introduction to the construction of combinational and sequential DesignWare parts. It does not take the place of the DesignWare Developer documentation, which supplies full details of the syntax and semantics of the various files. Instead, it will give you an overview of the files, requirements, and steps necessary to get your design running in simple cases.

## A.1  Combinational DesignWare

Combinational DesignWare is present in most designs whether you explicitly invoke it or not. For example, an instance of the VHDL operator '+' is mapped to an abstract operator that represents a signed or unsigned addition operator, depending on the types of its arguments and its result. The operator is then mapped in turn onto a DesignWare adder module, and ultimately to a particular implementation: e.g. a ripple-carry adder. This is normally almost invisible to the casual user of HDL Compiler and BC; but sometimes it is useful to be able to define your own operators. A very good example of this is given in Chapter 10. In that example, there is a moderately complex logical operation, the **sbox**, that has no special lexical symbol in either VHDL or Verilog. Instead, we define a function.

If you use the **preserve_function** pragma on the new function, the functionality you will get is very similar to that of a combinational DesignWare part. That is, the function can be scheduled and allocated, and the functional unit is shared.

A combinational DesignWare part, however, has some significant advantages over a function that is preserved.

First, combinational DesignWare can be parameterized. This means that you don't have to define the same function more than once for different bit widths. For example, suppose you needed a multiply/accumulate function. You might want to construct it only once, using unconstrained subtypes.

Second, a DesignWare part can have multiple implementations; the best implementation is automatically selected during logic compilation.

Third, you can map a DesignWare part onto an existing macrocell, and so capture components and structures for which you might not have descriptions.

Finally, you can license and sell DesignWare as intellectual property.

202

The pathway by which you create a DesignWare part consists of eight steps.

First, create a Verilog or VHDL function, task, or procedure that expresses the behavior that you want to capture. From here on, I will just refer to it as a *subprogram*, and explicitly say what kind of subprogram where there are differences in the way different kinds of subprogram are treated. The subprogram you create will most commonly be embedded in your behavioral HDL input. You could use a dummy process as well, but it's handy to be able to test the subprogram where it will be used.

Second, annotate the function with a **map_to_operator** pseudo-comment. The name given to the function in this step will be used later on to denote the Design-Ware operator that are constructing. In Verilog, the annotation is

```
// synopsys map_to_operator opname
```

In VHDL, the pseudo-comment is

```
-- synopsys map_to_operator opname
```

These annotations are to be placed immediately after the interface declarations of the subprogram. If the subprogram is specifically a function, then you need to add a pragma that names the output port of the DesignWare part. The output pragma should follow the mapping pragma.

```
// synopsys return_port_name Z
```

```
-- synopsys return_port_name Z
```

A secondary function of the **map_to_operator** pragma is to prevent synthesis of the subprogram as it is written in the original HDL text. However, we could not just leave the function definition blank there, because then the function would not simulate correctly. By leaving the function present in the original text, we keep our ability to simulate the function; by adding the pragma we redirect synthesis from the original text to our new DesignWare part.

The third step is to analyze the HDL text. This is done using dc_shell or bc_shell and the **analyze** command. You should check the WORK directory you are using after this step; there should be a pair of `.syn` files whose prefixes are those of the entity and architecture. In Verilog, the `.syn` file will be named after the module.

The fourth step is to run a utility Synopsys distributes as part of the DesignWare Developer product: the name of the utility is **create_synlib_template**.

```
unix> create_synlib_template design_name func_name
```

This utility will read in the `.syn` files alluded to above and produce a VHDL or Verilog wrapper file and a synthetic library (`.sl`) file. The `.sl` file describes the interface of the new component to the rest of a synthesized circuit. Henceforth I'll call this new file the 'DesignWare HDL'.

You need to pass **create_synlib_template** the names of the design and of the function to be encapsulated. You are permitted to pass it a number of other parameters; these will be listed if you try to invoke it with no arguments at all.

It is important that the file `.synopsys_vss_setup` points to the correct WORK directory when you run the `create_synlib_template` utility; otherwise the utility cannot find the right location of WORK and so cannot locate the `.syn` files.

The fifth step is to fill in the actual contents of the DesignWare HDL. The contents in question are an entity/architecture pair or a module that describes the pinout and behavior of the new DesignWare part. You will have to get the port bitwidths right, the entity/architecture names right (in Verilog, the module name must be set properly), and you will have to insert the function body into the new module as a process.

The other thing you have to do to the DesignWare HDL file is copy and paste the subprogram body from the original HDL text into the new HDL file. You will probably have to do some repair work as well: the generic bit widths will not figure in the original definition, and must be used correctly.

If you are using Verilog, the module name should be of the form `ee_aa`, where `ee` is the processor name and `aa` is the implementation name. In VHDL, the entity name will be mapped to the processor, and the architecture name will be mapped to the implementation.

The sixth step is to create the synthetic library database (`.sldb`) file. This is done by invoking dc_shell and issuing the commands

```
dc_shell> read_lib filename.sl
dc_shell> write_lib filename.sldb
```

Now these files can be stored in another directory; this is a good idea, because it reduces the possibility of confusion. A typical directory organization for custom DesignWare is shown in Fig. A.1.

In the seventh step you should define a design library. The logical design library name you use is associated with a directory path by the `-path` flag, and the path should point to the directory where you want the machine-readable form of the new DesignWare port to reside.

```
bc_shell> define_design_library dw_lib -path /dw/lib
```

Finally, analyze the DesignWare HDL text into the new design library using the `analyze` command. You are now ready to elaborate the rest of your design, i.e. the design in which the new DesignWare part is invoked.

## Compiled DesignWare

The new DesignWare part can also be compiled and dont_touched at this stage, in which case you will need to perform the following additional steps:

First, compile the HDL description of the DesignWare part, using whatever constraints and flags you deem appropriate.

Second, `dont_touch` all of the cells in the compiled part. This will prevent the part from being recompiled where it is instantiated.

Third, write out the compiled part. You can do this in one of two ways. The first alternative is to write it to WORK, where it will take the form of a `.ldb` file

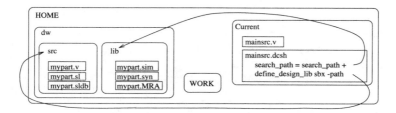

**Figure A.1.** Directory structure: custom DesignWare

that can be brought in automatically whenever you use that WORK directory. The second alternative is to write it out (using the **-h** flag) as a **.db** file, in which case you will have taken over responsibility for managing it, loading it in, updating it, and so on.

The compiled part can now be used as a macrocell by BC.

### Using the part

In order to use the new part, you should set the bc_shell **synthetic_library** variable to include the new synthetic library you have just created.

```
bc_shell> synthetic_library = synthetic_library + filename.sldb
```

You will also need to define the design library where the new part's machine readable files are located. In addition, you should set your search path to include the directory where the **.sldb** file is located. This is shown in Fig. A.1. In Fig. A.1 the directories are shown as rounded boxes, and files are shown as sharp-cornered boxes. The directory where your current project is located is **Current**; the directory where your custom DesignWare is located is **dw**. In **Current** is a source HDL file **myfile.v** and a script file **myfile.dcsh**. In **myfile.dcsh** I have shown a pair of lines schematically; these are the commands that define the design library, pointing to the **dw/lib** directory, and the one that sets the search path, pointing to the **dw/src** directory, respectively. The source code for the DesignWare part, the **.sl** and **.sldb** files are all located in the **src** directory; the machine-readable files are located in the **lib** directory.

### Example

The following example is taken from Chapter 10. In that chapter we needed a single function, the **sbox**, that was called many times; and we wanted to share it.

Here is the function definition. I have lifted it verbatim from the VHDL and Verilog in Chapter 10; see page 186 for the rest. First, the VHDL.

```
-- function describing the DES s boxes
function sbox (i: bit_vector (5 downto 0);
               n: bit_vector (2 downto 0))
     return bit_vector is
```

```vhdl
-- pragma return_port_name s
-- pragma map_to_operator sbox_op
variable indexU: unsigned(7 downto 0);
variable index: integer;
variable tbl: bit_vector(0 to 255);
begin
  indexU := (To_StdULogic(i(5)), To_StdULogic(i(0)),
             To_StdULogic(i(4)), To_StdULogic(i(3)),
             To_StdULogic(i(2)), To_StdULogic(i(1)),
             '0', '0');
  index := conv_integer(indexU);
  case n is
    when "000" => tbl := s1;
    when "001" => tbl := s2;
    when "010" => tbl := s3;
    when "011" => tbl := s4;
    when "100" => tbl := s5;
    when "101" => tbl := s6;
    when "110" => tbl := s7;
    when "111" => tbl := s8;
    when others => tbl := (others => '0');
  end case;
  return(tbl(index), tbl(index+1), tbl(index+2), tbl(index+3));
end sbox;
```

This is the Verilog version of the same function.

```verilog
`include "smax.v" // include the long s-box constants
module des (...)

... // irrelevant stuff. See Chapter 10

  // function descibing the DES s boxes
  function [3:0] sbox;
    // synopsys return_port_name s
    // synopsys map_to_operator sbox_op
    input [5:0] i;
    input [2:0] n;
    reg [7:0] index;
    reg [0:255] tbl;
    begin
      index = { i[5], i[0], i[4:1], 2'b00 };
      case (n)
        0: tbl = `s1;
        1: tbl = `s2;
```

```
           2:  tbl = 's3;
           3:  tbl = 's4;
           4:  tbl = 's5;
           5:  tbl = 's6;
           6:  tbl = 's7;
           7:  tbl = 's8;
         endcase
         sbox = {tbl[index],tbl[index+1],tbl[index+2],tbl[index+3]};
       end
     endfunction
     ...

 endmodule
```

Notice that the function of the sbox has been encapsulated in a single function that is called elsewhere. The pragmas **return_port_name** and **map_to_operator** have been added to the function.

First, analyze the original source file **des.v**.

```
         bc_shell> analyze -f verilog des.v
```

The result is two files located in the current WORK directory.

The next step is to run the program **create_synlib_template** with arguments that point it to the design and the function.

```
         unix> create_synlib_template des sbox
```

The result of this is two files: one called **DWSL_sbox_mod.v** and one called **my_dw.sl**. I could have gotten them into particular library and source directories by using optional flags; instead, I preferred to move them into the **dw** directories by hand. The **.v** file is renamed **sbox.v** and the **.sl** file is renamed **sbx.sl** and both are moved into /dw/src.

```
       unix> mkdir ~/dw
       unix> mkdir ~/dw/src
       unix> mkdir ~/dw/lib
       unix> mv DWSL_sbox_mod.v ~/dw/src/sbox.v
       unix> mv my_dw.sl ~/dw/src/sbx.sl
```

The next step is to edit the new file **sbox.v** so that the names are right and the template contains the proper functionality.

1. Change the name of the module in the file **sbox.v** to be **sbox_rtl**. In VHDL, change the entity name to be **sbox** and the architecture name to be **rtl**. These names are important, as they are used by BC to find the processor and the architecture later on.

2. Set the parameters that describe the bit widths of the ports of the DesignWare part. Alternatively, leave them alone; then the **.sl** file needs hdl_parameter

lines, as described in the DesignWare Developer manual. Declare the output
variable **s** to be a **reg** if you are using Verilog.

3. Create an RTL process beginning with **always @(i or n)** (in VHDL, the
   sensitivity list consists of **i** and **n**. and including the function body of the
   original function.

Here is the Verilog text of the new function.

```verilog
`include "smax.v"
module sbox__rtl(i, n, s);
  // synopsys dc_script_begin
  // set_model_load 4 all_outputs()
  // set_model_drive 1 all_inputs()
  // max_area 0
  // synopsys dc_script_end
  parameter i_width = 6;
  parameter n_width = 3;
  parameter s_width = 4;
  input [i_width-1:0] i;
  input [n_width-1:0] n;
  output [s_width-1:0] s;
  reg [s_width-1:0] s;
  reg [7:0] index;
  reg [0:255] tbl;
  always @ (i or n) begin
      index = { i[5], i[0], i[4:1], 2'b00 };
      case (n)
        0: tbl = `s1;
        1: tbl = `s2;
        2: tbl = `s3;
        3: tbl = `s4;
        4: tbl = `s5;
        5: tbl = `s6;
        6: tbl = `s7;
        7: tbl = `s8;
      endcase
      s <= {tbl[index],tbl[index+1],tbl[index+2],tbl[index+3]};
  end
endmodule
```

The changes in VHDL are analogous; you must define a new process with the proper
sensitivity list, and returning the correct result.

The next step is to analyze the new file into the  `/dw/lib` directory.

```
bc_shell> define_design_lib sbx -path ../lib
bc_shell> analyze -f verilog sbox.v -work sbx
```

Next we must edit the sbx.sl file to make the names correspond to one another correctly.

1. Change the name of the library declaration to **sbx.sldb**.

2. Change the name of the module to **sbox**.

3. Change the name of the **design_library** to **sbx**.

Now build the **sbx.sldb** synthetic library data base file:

```
bc_shell> read_lib sbx.sl
bc_shell> write_lib sbx.sldb -o sbx.sldb
```

The next thing to do is to go back to your original directory and edit the script you will be using to schedule the original module. The important thing here is to set up the links as shown in Fig. A.1.

```
bc_shell> define_design_lib sbx -path ~/dw/lib
bc_shell> search_path = search_path + {~/dw/src}
bc_shell> synthetic_library = synthetic_library + sbx.sldb
```

You should now be able to run the entire script with no further changes and no error messages. Fig. A.2 shows the various files schematically, with the name correspondences indicated by dashed arrows. Solid arrows indicate the ancestry of the files; I have marked some of the solid arrows to indicate the name of the compilation command.

Note that any of these name correspondences can be wrong for a variety of reasons. If this is the case the whole flow will probably fail; you will find it necessary to scan for some disagreement between the names of two or more of the objects.

## A.2  Sequential DesignWare

In this Section we will use sequential DesignWare to construct wrappers for a simple RAM. The example is taken from Chapter 9. The same techniques that are used here can also be used to construct your own sequential components to perform other arithmetic and logical functions; the difference is that for a general sequential part you will have to provide a synthesizable VHDL description, which is not the case with RAMs.

A *RAM wrapper* is a construct that hides the timing details of a vendor's asynchronous or pseudosynchronous RAM. It provides a clean, synchronous model that BC can interface into the rest of the circuit.

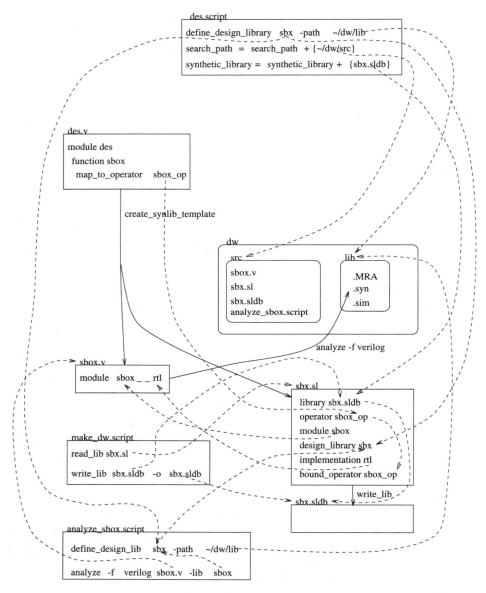

**Figure A.2.** Name correspondences: DesignWare

You can get the RAM wrapper in one of two ways. First, your silicon vendor can provide a RAM wrapper tuned to the particular RAMs the vendor provides. This is useful, but many vendors appear to be waiting for customer demand to justify internal development of RAM wrappers.

Second, you can build your own. This means constructing your own synchronous RAM interfaces for the RAMs your vendor provides. I cannot advise you on the circuitry of the interfaces: it depends too much on the individual RAM you are trying to wrap. Generally speaking, the function of the wrapper is to take an asynchronous or pseudosynchronous RAM macro, as provided by the vendor, and give it edge-triggered, single-clock interface timing. Thus the RAM will be made to look, for timing purposes, like a shift register with some small number of stages.

In the wider sense, the RAM wrapper also provides the appropriate apparatus to tell BC how to use the RAM. This is basically a machine-readable timing diagram for the synchronized part. For example, the wrapped RAM might need to see a write enable in the first clock cycle of a write; a valid address in the second; and valid data in the third. BC needs this interfacing information in order to be able to properly control and connect the wrapper (and hence the RAM) to the rest of the synthesized circuit.

The entire wrapper, in the largest sense, is described by four files. For concreteness' sake I will use the name **myrams** as a prefix for the custom wrappers I am about to describe.

These files can be organized in a directory we'll call **ram**; I will describe the directory as if it is located immediately inside your home directory, i.e. in the forthcoming example I'll call it ˜**/ram**. You will need some or all of the following files, depending on how much of the wrapper you have to construct yourself and how much you get from the vendor.

**myrams.sldb** is in the best case provided by the vendor as part of the wrapper. If not, you will have to construct it using DesignWare Developer and another file called **myrams.sl**. DesignWare Developer compiles the **.sl** file to get the **.sldb** file. The **.sldb** file is a machine-readable description of the access protocol of the RAM; e.g. it might state that in the second cycle of a write access the address lines should be asserted. BC uses this information to connect up the RAM and generate the proper control signals when you use the RAM in a circuit. The **.sldb** file represents the timing diagram of the RAM as seen from outside the wrapper, i.e. from BC's point of view.

**myrams.db** is a technology library file, which is a compiled (machine-readable) form of a file **myrams.lib**. If you are getting your wrappers from a vendor, the vendor is responsible for processing the **.lib** file, using Library Compiler, to get the corresponding **.db** file; you would never see the **.lib** file at all. The function of the **.db** file is to provide a technology mapping database for the RAM; for example, the area, timing, and parasitic parameters of the RAM are detailed here. If the vendor has not provided the wrapper, there should still be a **.db** file that describes the asynchronous RAM macro. If all else fails, you can fake it using Library Compiler; that's the way I did it in the Chapter 9. But sooner or later you will need the real RAM macro and that has to come from the vendor.

**ram32x24.vhd** This file is the VHDL description of the wrapper for a particular RAM. Note that you can use a VHDL wrapper even if your site uses Verilog. These files (and in general there might be many, as you use a generator to provide yourself with many different sizes and shapes of RAM) are in the best case provided by the vendor. The `.vhd` file instantiates the asynchronous or pseudosynchronous RAM as a `component` inside an `entity`; the rest of the `entity` is the synchronizing logic (registers, gates, etc.) needed to convert the RAM macro to a synchronous part. The `entity` declaration's `port` declaration part gives the pinout that is referenced in the `.sl` file's description of the synchronous timing diagram. The architecture body of the entity contains a `component`, which is an instantiation of the asynchronous or pseudosynchronous RAM macro that the `.lib` file describes, and hence is a reference to a cell in the `myrams.db` file. The architecture body also contains registers, gates, etc. as needed to adapt and synchronize the macro and make the entity as a whole fully synchronous.

**myrams.scr** This is a bc_shell script that compiles the files listed above. Normally you would not see this file because the vendor would be responsible for compiling and packaging the libraries for you. I will provide an example so that you can build your own as necessity dictates.

### Example

Suppose we want to design a circuit using a 32-word RAM having a 24-bit word width. Suppose further that we don't have any support at all from the vendor; perhaps we have not yet decided on the vendor or the technology, but we want to get a simulatable and synthesizable version of our circuit anyway.

We begin by constructing the file `myrams.lib`. This, recall, is a technology mapping file that we will compile into a `.db` file using Library Compiler.

```
library (myrams) {
   type (bus5) {
      base_type : array;
      data_type : bit;
      bit_width : 5;
      bit_from : 4;
      bit_to : 0;
      downto : true;
   }
```

```
type (bus24) {
   base_type : array;
   data_type : bit;
   bit_width : 24;
   bit_from : 23;
   bit_to : 0;
   downto : true;
}
cell(ram32x24) {
   area : 16000;
   pin(we){
      direction : input;
      capacitance : 0.0 ;
   }
   pin(oe){
      direction : input;
      capacitance : 0.0 ;
   }
   bus(di){
      bus_type : bus24
      direction : input;
      capacitance : 0.0 ;
   }
   bus(adr){
      bus_type : bus5
      direction : input;
      capacitance : 0.0 ;
   }
   pin(clk){
      direction : input;
      capacitance : 0.0 ;
   }
   bus(do){
      bus_type : bus24
      direction : output;
      capacitance : 0.0 ;
   }
}
}
```

This file defines a **library** named *myrams*. The library consists of two bus type declarations and a single cell definition.

The bus type declarations associate a name (e.g. **bus5**) with a base type, which is **array**, the type of the elements, a bit width, and the indexing scheme of the bus. The declaration of **bus5** is therefore equivalent to something like

```
type bus5 is array (4 downto 0) of bit;
```

We need the definition of the five-bit bus for the address lines of the 32-word memory, and the 26-bit bus for the data lines.

The cell definition gives us the pinout of and some other technology information about the macro. Because this is a dummy macro that I'm using just for the example of Chapter 9, I am providing some very abstract data on the technology. For example, I arbitrarily decided to make the area of the macro 16000 units.

The pinout consists of six pins. The pins are classified as to direction (input or output) and capacitance. In addition, if the pins are array pins you need the bus types as well. Thus the pin *we* is a one-bit input with a capacitive loading of 0 units, and *do* is a 26-bit output with 0 units of capacitance.

Additional fields (and better numbers!) will make the macro more realistic:

**units** for time, voltage, current, capacitive load, pulling resistance, and nominal environmental parameters can be defined at the library level;

**operating conditions** can be defined, again at the library level, as combinations of temperature, voltage, process and delay models;

**wire loads** can be defined as combinations of resistance, capacitance, and area;

**pin timing** can be defined in terms of rise and fall times, resistances, slopes, and the related pins that cause an output pin to transition.

The next file is the VHDL description of the wrapper.

```
library ieee;
use ieee.std_logic_1164.all;
library GTECH;
use GTECH.GTECH_components.all;

entity my32x24ram is
   generic (addr_width : INTEGER :=  5;
            data_width : INTEGER :=  24);
   port(  DATAO  : out std_logic_vector
                       (data_width - 1 downto 0);
          ADDR   : in std_logic_vector
                       (addr_width - 1 downto 0);
          DATAI  : in std_logic_vector
                       (data_width - 1 downto 0);
          ENABLE : in std_logic;
          CLK    : in std_logic
          );
```

```vhdl
end my32x24ram;
architecture wrap of my32x24ram is
   -- pragma dc_script_begin
   -- set_local_link_library myrams.db
   -- set_dont_touch current_design
   -- pragma dc_script_end
   component ram32x24
   port ( we   : in  std_logic;
          oe   : in  std_logic;
          di   : in  std_logic_vector(data_width-1 downto 0);
          adr  : in  std_logic_vector(addr_width-1 downto 0);
          clk  : in  std_logic;
          do   : out std_logic_vector(data_width-1 downto 0)
          );
   end component;

   Signal CONSTANT_LOGIC_0     : std_logic;
   Signal CONSTANT_LOGIC_1     : std_logic;
   Signal CLK_INV              : std_logic;
   Signal DATA_OUT_HOLD        : std_logic_vector
                                   (data_width-1 downto 0);

   begin
      CONSTANT_LOGIC_1 <= '1';
      CLK_INV          <= not CLK;

   instram : ram32x24
   port map ( we    => ENABLE,
              oe    => CONSTANT_LOGIC_1,
              di    => DATAI,
              adr   => ADDR,
              clk   => CLK_INV,
              do    => DATA_OUT_HOLD
             );
   -- Register the output of the RAM wrapper
   -- at the positive edge of the clock.
   -- This ensures that the outputs from the RAM are stable
   -- before going out to the datapath.
   f4: for i in data_width - 1 downto 0 generate
      flp_data: GTECH_FD1
         port map (DATA_OUT_HOLD1(i), CLK, DATAO(i), open);
   end generate;
end wrap;
```

The VHDL file begins with the declaration of the **entity** that represents the wrapped RAM. I've called this entity 'my32x24ram'. The entity has the pinout of the RAM we want; in this case it is a single-port RAM with a write enable, an output enable, and separate data input and output pins. It is important that the clock pin of all sequential parts be named CLK and nothing else.

The architecture of the entity contains the component declaration of **ram32x24**, which is a reference to the macro defined in the **.lib** file. The names must match. The instantiated component is called **instram**; it is wired so that the output enable is always true, the clock is inverted, and the data output line drives a bank of flip-flops (i.e. a register) that holds the output stable and thus synchronizes the output. For a more complex RAM macro, or one with uglier timing characteristics, the VHDL wrapper circuit would have to be designed accordingly.

You should take particular notice of the pragma comments at the top of the VHDL. These will be interpreted as a partial script that tells the Design Compiler how to process and link up the wrapper design.

The next file we will need is the file **myrams.sl**. This file describes the protocols by which the synchronized RAM is to be accessed. It uses the pinout defined in the wrapper VHDL and gives BC instructions on how to access and control the RAM in the synthesized circuit.

```
library (myrams.sldb) {
  module (my32x24ram) {
    design_library : "rams";
    parameter(addr_width) {
      formula : "width('ADDR')";
      hdl_parameter : true;
    }
    parameter(data_width) {
      formula : "width('DATAI')";
      hdl_parameter : true;
    }
    clocking_scheme() {
      clock_type : positive_edge;
    }
    pin (DATAO) {
      direction : output;
      bit_width : "data_width";
      stall : active_low;
    }
    pin (ADDR) {
      direction : input;
      bit_width : "addr_width";
      stall : active_low;
    }
```

```
  pin (DATAI) {
    direction : input;
    bit_width : "data_width";
    stall : active_low;
  }
  pin (ENABLE) {
    direction : input;
    bit_width : "1";
    stall : active_high;
  }
  pin (CLK) {
    direction : input;
    bit_width : "1";
    clock_pin : true;
  }
  implementation("wrap") { }
  resource (S1) { count: n; }
  resource (S2) {}
/*    Sample timing looks like :
      | Apply Address |  Data valid  | */
  binding(read) {
    bound_operator : "MEM_READ_SEQ_OP";
    state() {
      pin_association(ADDR) {oper_pin : ADDR;}
      pin_association(CLK) {oper_pin : CLK;}
      unbound_oper_pin(CLK) { value : "1"; }
      use_resource(S1) {}
    }
    state() {
      pin_association(DATAO) {oper_pin : Q; stable : true;}
      pin_association(CLK) {oper_pin : CLK;}
      unbound_oper_pin(CLK) { value : "1"; }
      use_resource(S2) {}
    }
  }
  binding(write) {
    bound_operator : "MEM_WRITE_SEQ_OP";
/* The write cycle will look like | ADDR/DATA/ENABLE | */
    state() {
      pin_association(ADDR) {oper_pin : ADDR;}
      pin_association(DATAI) {oper_pin : D;}
      pin_association(ENABLE) {value : "1";}
      pin_association(CLK) {oper_pin : CLK;}
      unbound_oper_pin(CLK) { value : "1"; }
```

```
        use_resource(S1) {}
    }
  }
}
```

The `.sl` file defines a synthetic **library**, here called **myrams.sldb**. The library consists of one or more modules, each module being the protocol description of a single type of functional unit. Shown here is the module describing the entity **my32x24ram**. This module is part of a logical library named **rams**.

The module can be parameterized, here with two parameters: the address and the data bit widths. These parameters are taken from the HDL description, and they are computed as being the width of the address and data pins respectively. Other formulas can be used here instead of **width**; you should see the DesignWare Developer documentation for a complete listing of the syntax and semantics of its computational abilities. The parameters themselves do not mean anything: they are associations of integers with parameter names. The parameters names can be referenced later in the module definition.

The next field is the clocking scheme. This describes the way the sequential part is clocked, here specified as being on positive edges. Notice that you would have to change this if your BC circuit used a negative-edge clock; BC does not automatically invert the clock.

The next fields give the pinout of the wrapped RAM. These are characterized in terms of direction (input or output), pin bitwidth (which may refer to parameters or be constants), and a **stall** attribute that tells you what the state of the input pin should be when the component is not being used.

The **implementation** of the module corresponds to an **architecture** in the VHDL file. If you define multiple architectures, then BC can in theory select one on the basis of timing, area, or other criteria; but at the time of writing this feature has not yet been implemented except for combinational DesignWare, which has implementations selected during logic compilation.

The implementation of a sequential part defines one or more **resources**. These are abstract representations of internal blocks of the sequential part; they are declared as being **use**d during particular operations. Each resource may also have an optional **count**, which tells how many of that resource are available. If a resource is available during a particular cstep, then the corresponding operation can be performed; if not, then the operation cannot and BC will have to schedule the operation in another cstep or allocate another DesignWare module.

The implementation also defines a collection of **bindings**. These are used by BC to attach a logical operation to the component implementation. Each binding consists of two parts.

First, a **bound_operator** tells BC the name of an operator that can be bound to the implementation. Familiar examples in the combinational world are **ADD_UNS_OP** and **MUL_TC_OP**; in the area of RAMs, a typical operation is **MEM_READ_SEQ_OP**. These operator names are generated by the HDL elaboration process, and they represent the abstract additions, multiplications, and memory accesses inferred by BC from lexical constructs such as +, *, and [].

The second part of the binding is an ordered set of **states**. Each state consists of constraints, unbound pins, pin associations, and resources used; where a constraint allows a control vector to be applied, the unbound pins allow pins to be declared that are not used in the abstract operator, the pin association associates a pin of the implementation with a formal parameter or result of the abstract operator, and the **use_resource** field describes how many of each type of the declared resources will be used during the cstep in which the state is valid.

Thus in the example above, the first binding is named **read**; it is bound to the abstract operator **MEM_READ_SEQ_OP**. The read itself is a two-state operation. The first statement of the first state tells BC that the address pin is to be driven with the logical address that drives the abstract operator. The second statement tells BC that the clock pin is to be wired to the clock of the circuit being synthesized; the **unbound_oper_pin** declaration of the clock tells BC that no particular value is to be associated with the clock. This allows the clock pin to be driven by the clock net of the synthesized design. Finally, the state is completed with a list of the resources consumed. Here the resource **S1** is consumed; this resource corresponds to the macro's internal hardware that decodes addresses and decides whether or not the operation is a write.

Be very careful with your declarations and uses of resources: in this example, because the resource **S1** is *not* used in the second cycle of the read, BC can schedule the first cycle of a read operation concurrently with the second cycle of an earlier read operation; that is, it is possible to use this part for pipelined reads. If there was no internal address register, etc. in the RAM, then the address lines could not be allowed to change after the first cycle, and this description would have to be modified so that **S1** would also be used in the second cycle, thus preventing concurrent scheduling of two phases of two different read operations.

Pins can also be declared to be **stable** in a state. This has different meanings for input and output pins. If an input pin is stable in a state, then BC must register the input, as in a multicycle operation. If an output is stable in a state, then the output is registered internally, and BC does not have to store the result until the next operation accesses that pin. This example shows a stable output pin; referring to the VHDL wrapper, the stabilizing register is the result of the final **generate** statement of the architecture body.

Note that both the **stable** attribute and resource consumption must be declared under some circumstances: i.e. when the input must be registered and held over multiple cycles by circuitry external to the wrapped RAM.

In the example RAM the read operation takes two cycles, and reads can be pipelined; the write operation takes one cycle, and a write can be scheduled concurrently with the second step of a read. Notice, however, that a write cannot be scheduled concurrently with the first step of a read, because they consume the same resource **S1**, and there is only one **S1** available.

The next file you need, if you are constructing your own wrappers, is the file `myrams.scr`. This is a script file that will invoke DesignWare developer, which will then read in the other files and create the `.sldb` and `.db` files.

```
analyze -f vhdl ram32x24.vhd
read_lib myrams.lib
write_lib myrams -o myrams.db
read_lib myrams.sl
write_lib myrams.sldb -o myrams.sldb
quit
```

This script begins by reading in the VHDL file. It then reads in the `.lib` file and writes out the `.db` file. If you already have the `.db` file, you don't need to process the `.lib` file. The next thing that happens is that the `.sl` file is read, and the `.sldb` file constructed. Once you have both the `.db` and `.sldb` files, you have your wrapped RAMs.

Now all that is necessary is to set the synthetic library list, search path, and design library in your main synthesis script (e.g. `idct.dc` in Chapter 9).

```
synthetic_library = synthetic_library + myrams.sldb
search_path = search_path + ~/rams
/* schedule */
define_design_lib rams -path ~/rams
/* compile */
```

# Appendix B

# Synthesizable Subsets

This appendix gives a brief overview of the synthesizable subsets of VHDL and Verilog. For those of us who are already familiar with the RTL synthesizable subset, the BC subset is essentially the same, with the exception that BC does not support tristate logic and RTL does not support the transport delay BC uses for pipelining loops in fixed mode.

The pragmas synthesis_on, synthesis_off, translate_on, and translate_off are supported in both languages. In VHDL pragmas can be begun with either pragma or synopsys; in Verilog pragmas can be begun with synopsys only.

## B.1  VHDL Subset

The constructs that BC fully supports are those that can be compiled into a circuit. Constructs that cannot be supported at all, or that are ignored, are those that make sense in a simulation context but that do not make sense when mapped directly to hardware.

Fully supported constructs are:

1. Arithmetic, logical, and relational operators. Division is supported by an IEEE package.

2. Architecture bodies.

3. Arrays

4. Array variable assignment.

5. The following attributes: RIGHT, LEFT, HIGH, LOW, BASE, RANGE, LENGTH.

6. Component declarations and instantiations.

7. Concurrent procedure calls.

8. Concurrent signal assignments.

9. Constant declarations.

10. Entity declarations.

11. Enumerated types.

12. Function calls.

13. Generate statements.

14. If, then, else, elsif, case statements.

15. Indexed and slice names.

16. Integer declarations.

17. Integer types.

18. Libraries.

19. Loop statements.

20. Next and return statements.

21. Null statement.

22. Operator overloading.

23. Package declarations and bodies.

24. Procedure call statements.

25. Process statements.

26. Qualified and static expressions.

27. Records.

28. Signal, variable, and interface declarations.

29. Simple names.

30. Subprogram declarations and bodies.

31. Subprogram overloading.

32. Type and subtype declarations.

33. Type conversions.

## B.1.1 VHDL constructs that are only partially supported

1. Aggregates are supported in most cases.

2. Attribute names, declarations and specifications. Some attributes don't make sense or they have no electrical counterparts. These are listed below.

3. EVENT is supported in the form of clock statements.

4. Exit statements are supported if they are reset exits or local loop exits.

5. Resolution functions are supported only for wired and, or, and three-state.

6. Selected names are supported unless they go outside the current design.

7. Signal and variable assignment statements are supported as long as it isn't a waveform being assigned.

8. STANDARD is mostly supported. But some of the constructs e.g. the attributes are not supported.

9. Transport delays are supported in BC as part of the pipelined

10. Wait statements are supported when they are clock edge waits.

## B.1.2 VHDL constructs that are ignored by synthesis

1. Access and file types, file declarations don't make sense in synthesis.

2. Aliases.

3. Assertions don't make sense in synthesis.

4. Configuration specifications and declarations.

5. Floating point types.

6. Incomplete type declarations.

7. Physical types don't make sense in synthesis.

## B.1.3 Unsupported VHDL Constructs

1. Allocators make no sense in synthesis.

2. Attributes POS, VAL, SUCC, PRED, LEFTOF, RIGHTOF, DELAYED, TRANSACTION, RIGHT(N), LEFT(N), RANGE(N), ACTIVE, LAST_EVENT, LAST_ACTIVE, REVERSE_RANGE(N), HIGH(N), LOW(N), LAST_VALUE, BEHAVIOR, STRUCTURE.

3. Context information.

4. Disconnections.

5. Short-Circuit operators

6. TEXTIO

## B.1.4   Special attributes for synthesis in VHDL

In VHDL it is possible to put attributes on many constructs. Synthesis supports some special VHDL attributes; mostly these have counterparts in the bc_shell command language as well. In Verilog, you have to use the command language counterparts.

1. ARRIVAL, FALL_ARRIVAL, RISE_ARRIVAL

2. DRIVE, RISE_DRIVE, FALL_DRIVE

3. LOGIC_ONE, LOGIC_ZERO

4. EQUAL, OPPOSITE

5. DONT_TOUCH_NETWORK

6. LOAD

7. DONT_TOUCH

8. MAX_AREA

9. ENUM_ENCODING

10. MAX_TRANSITION, MAX_DELAY, MAX_RISE_DELAY, MAX_FALL_DE-LAY, MIN_DELAY, MIN_RISE_DELAY, MIN_FALL_DELAY,

11. UNCONNECTED

12. HOLD_CHECK, SETUP_CHECK

## B.2   Verilog Subset

## B.2.1   Fully supported Verilog constructs

1. Always blocks.

2. Arithmetic and logical operators other than the triple operators (=== and !==).

3. Case, casex, casez statements

4. 'Define

5. For loops. Note that the loop test can only be >, <, >=, or <=.

6. Function definitions and calls.

7. 'Include.

8. Memory declarations.

9. Modules and macromodules. Note however that inlining of macromodules, which is done in simulation, is not done in synthesis. Thus modules and macromodules are treated in the same way.

10. Parameters.

11. Reg and wire declarations.

12. Tasks.

13. Technology-independent logical functions, e.g. and, or, not.

14. Wand, wor, wire.

## B.2.2   Verilog constructs that are not fully supported

1. @ is supported in posedge and negedge clock statements.

2. Delays are supported in BC as part of the pipelined loop methodology.

3. Disable of local loops and reset loops are supported by BC; all disables are supported by HDLC.

4. Division and modulus are allowed for powers of two only

5. Hierarchical names within a module are supported if they don't escape from the module or create new ports.

6. Ranges and arrays for integers are supported for some choices.

## B.2.3   Verilog constructs that are not supported

The following Verilog constructs are not supported by synthesis:

1. Bit-selected left hand side of assignments with a variable as index; would result in combinational feedback in some cases.

2. Defparam isn't supported; an equivalent is supported.

3. Deassign

4. Event doesn't make sense in synthesis.

5. Force, release don't make sense in synthesis.

6. Fork

7. 'Ifdef, 'endif, and 'else

8. Initial doesn't make sense in synthesis.

9. Nmos, pmos, etc. are analog constructs.

10. Primitives

11. Simulation and system directives don't make sense in synthesis.

12. Times and events

13. Triand, trior, trizero, trireg, pullup, tranif0, rtran etc. are analog constructs.

14. Wait.

# Bibliography

[1] R. A. Bergamaschi, A. Kuehlmann, S.-M. Wu, V. Venkataraman, D. Reischauer, and D Neumann. A methodology for the production use of high level synthesis. In *Proceedings of the 6th International Workshop on Hight Level Synthesis*. IEEE, 1992.

[2] G. Berry and G. Gonthier. The ESTEREL synchronous programming language: Design, semantics, implementation. *Science of Computer Programming*, 19:87–152, 1992.

[3] Eli Biham and Adi Shamir. *Differential Cryptanalysis of the Data Encryption Standard*. Springer-Verlag, 1993.

[4] Raul Camposano. Path based scheduling for synthesis. *IEEE Transactions on CAD*, 1(5):85–93, 1991.

[5] Daniel D. Gajski, Nikil D. Dutt, Allen C. H. Wu, and Steve Y. L. Lin. *High-Level Synthesis: Introduction to Chip and System Design*. Kluwer, 1992.

[6] Catherine H. Gebotys and Mohamed I. Elmasry. *Optimal VLSI Architecture Synthesis: Area, Performance, and Testability*. Kluwer, 1992.

[7] D. Harel. Statecharts: A visual formalism for complex systems. *Science of Computer Programming*, (8), 1987.

[8] O. J. Joeressen, M. Oerder, R. Serra, and H. Meyr. DIRECS: System design of a 100Mbit/s digital receiver. *IEE Proceedings-G*, 139(2):222–230, April 1992.

[9] T. Kim and C. L. Liu. An integrated data path synthesis algorithm based on network flow method. In *Proceedings of CICC 95*. IEEE, 1995.

[10] J. Kunkel. COSSAP: A stream driven simulator. In *IEEE International Workshop on Microelectronics in Communications, Interlaken, Switzerland*. March 1991.

[11] Tai A. Ly, David W. Knapp, Ron Miller, and Don MacMillen. Scheduling using behavioral templates. In *Proceedings of the 32nd ACM/IEEE Design Automation Conference*, pages 101–106. IEEE, 1995.

[12] Tai A. Ly and Jack T. Mowchenko. Applying simulated evolution to high level synthesis. *IEEE Transactions on CAD*, 12(3):389–409, March 1993.

[13] Petra Michel, Ulrich Lauther, and Peter Duzy. *The Synthesis Approach to Digital System Design*. Kluwer, 1992.

[14] G. De Micheli. *Synthesis and Optimization of Digital Circuits*. McGraw-Hill, 1994.

[15] G. De Micheli, D. Ku, F Mailhot, and T. Truong. The Olympus synthesis system for digital design. *IEEE Design and Test*, pages 37–53, October 1990.

[16] Y. Nakamura, K. Oguri, and A. Nagoya. Synthesis from pure behavioral descriptions. In R. Camposano and W. Wolf, editors, *High-Level VLSI Synthesis*. Kluwer, 1991.

[17] National Bureau of Standards. *Data Encryption Standard*. National Bureau of Standards, January 1977. FIPS publication 46.

[18] Pierre G. Paulin and John P. Knight. Force directed scheduling in automatic data path synthesis. In *Proceedings of the 24th ACM/IEEE Design Automation Conference*. IEEE, 1987.

[19] W. H. Press, S. A. Teukolsky, W. T. Vettering, and B. P. Flannery. *Numerical Recipes in C: the Art of Scientific Computing*. Cambridge University Press, 1992.

[20] S. Ritz and H. J. Schlebusch. A parallel block diagram oriented simulator on a multicomputer system. In *3rd IEEE Workshop on CAMAD of Commun. Links and Networks*, Torino, Italy, Sept. 1990.

[21] D. Thomas, E. Lagnese, R. Walker, J. Nestor, J. Rajan, and R. Blackburn. *Algorithmic and Register-Transfer Level Synthesis: the System Architect's Workbench*. Kluwer, 1990.

[22] Jan Vanhoof, Karl Van Rompaey, Ivo Bolsens, Gert Goossens, and Hugo De Man. *High-Level Synthesis for Real-Time Digital Signal Processing*. Kluwer, 1993.

[23] Kazutoshi Wakabayashi. Cyber: High level synthesis system from software into ASIC. In R. Camposano and W. Wolf, editors, *High-Level Synthesis*. Kluwer, 1991.

# Index

## LICENSE AGREEMENT AND LIMITED WARRANTY

READ THE FOLLOWING TERMS AND CONDITIONS CAREFULLY BEFORE OPENING THIS DISK PACKAGE. THIS LEGAL DOCUMENT IS AN AGREEMENT BETWEEN YOU AND PRENTICE-HALL, INC. (THE "COMPANY"). BY OPENING THIS SEALED DISK PACKAGE, YOU ARE AGREEING TO BE BOUND BY THESE TERMS AND CONDITIONS. IF YOU DO NOT AGREE WITH THESE TERMS AND CONDITIONS, DO NOT OPEN THE DISK PACKAGE. PROMPTLY RETURN THE UNOPENED DISK PACKAGE AND ALL ACCOMPANYING ITEMS TO THE PLACE YOU OBTAINED THEM FOR A FULL REFUND OF ANY SUMS YOU HAVE PAID.

1.　　　**GRANT OF LICENSE:** In consideration of your payment of the license fee, which is part of the price you paid for this product, and your agreement to abide by the terms and conditions of this Agreement, the Company grants to you a nonexclusive right to use and display the copy of the enclosed software program (hereinafter the "SOFTWARE") on a single computer (i.e., with a single CPU) at a single location so long as you comply with the terms of this Agreement. The Company reserves all rights not expressly granted to you under this Agreement.

2.　　　**OWNERSHIP OF SOFTWARE:** You own only the magnetic or physical media (the enclosed disks) on which the SOFTWARE is recorded or fixed, but the Company retains all the rights, title, and ownership to the SOFTWARE recorded on the original disk copy(ies) and all subsequent copies of the SOFTWARE, regardless of the form or media on which the original or other copies may exist. This license is not a sale of the original SOFTWARE or any copy to you.

3.　　　**COPY RESTRICTIONS:** This SOFTWARE and the accompanying printed materials and user manual (the "Documentation") are the subject of copyright. You may not copy the Documentation or the SOFTWARE, except that you may make a single copy of the SOFTWARE for backup or archival purposes only. You may be held legally responsible for any copying or copyright infringement which is caused or encouraged by your failure to abide by the terms of this restriction.

4.　　　**USE RESTRICTIONS:** You may not network the SOFTWARE or otherwise use it on more than one computer or computer terminal at the same time. You may physically transfer the SOFTWARE from one computer to another provided that the SOFTWARE is used on only one computer at a time. You may not distribute copies of the SOFTWARE or Documentation to others. You may not reverse engineer, disassemble, decompile, modify, adapt, translate, or create derivative works based on the SOFTWARE or the Documentation without the prior written consent of the Company.

5.　　　**TRANSFER RESTRICTIONS:** The enclosed SOFTWARE is licensed only to you and may not be transferred to any one else without the prior written consent of the Company. Any unauthorized transfer of the SOFTWARE shall result in the immediate termination of this Agreement.

6.　　　**TERMINATION:** This license is effective until terminated. This license will terminate automatically without notice from the Company and become null and void if you fail to comply with any provisions or limitations of this license. Upon termination, you shall destroy the Documentation and all copies of the SOFTWARE. All provisions of this Agreement as to warranties, limitation of liability, remedies or damages, and our ownership rights shall survive termination.

7.　　　**MISCELLANEOUS:** This Agreement shall be construed in accordance with the laws of the United States of America and the State of New York and shall benefit the Company, its affiliates, and assignees.

8.　　　**LIMITED WARRANTY AND DISCLAIMER OF WARRANTY:** The Company warrants that the SOFTWARE, when properly used in accordance with the Documentation, will operate in substantial conformity with the description of the SOFTWARE set forth in the Documentation. The Company does not warrant that the SOFTWARE will meet your requirements or that the operation of the SOFTWARE will be uninterrupted or error-free. The Company warrants that the media on which the SOFTWARE is delivered shall be free from defects in materials and workmanship under normal use for a period of thirty (30) days from the date of your purchase. Your only remedy and the Company's only obligation under these limited warranties is, at the Company's option, return of the warranted item for a refund of any amounts paid by you or replacement of the item. Any replacement of SOFTWARE or media under the warranties shall not extend the original warranty period. The limited warranty set forth above shall not apply to any SOFTWARE which the Company determines in good faith has been subject to misuse, neglect, improper installation, repair, alteration, or damage by you. EXCEPT FOR THE EXPRESSED WARRANTIES SET FORTH ABOVE, THE COMPANY DISCLAIMS ALL WARRANTIES, EXPRESS OR IMPLIED, INCLUDING WITHOUT LIMITATION, THE IMPLIED WARRANTIES OF MERCHANTABILITY AND FITNESS FOR A PARTICULAR PURPOSE. EXCEPT FOR THE EXPRESS WARRANTY SET FORTH ABOVE, THE COMPANY DOES NOT WARRANT, GUARANTEE, OR MAKE ANY REPRESENTATION REGARDING THE USE OR THE RESULTS OF THE USE OF THE SOFTWARE IN TERMS OF ITS CORRECTNESS, ACCURACY, RELIABILITY, CURRENTNESS, OR OTHERWISE.

IN NO EVENT, SHALL THE COMPANY OR ITS EMPLOYEES, AGENTS, SUPPLIERS, OR CONTRACTORS BE LIABLE FOR ANY INCIDENTAL, INDIRECT, SPECIAL, OR CONSEQUENTIAL DAMAGES ARISING OUT OF OR IN CONNECTION WITH THE LICENSE GRANTED UNDER THIS AGREEMENT, OR FOR LOSS OF USE, LOSS OF DATA, LOSS OF INCOME OR PROFIT, OR OTHER LOSSES, SUSTAINED AS A RESULT OF INJURY TO ANY PERSON, OR LOSS OF OR DAMAGE TO PROPERTY, OR CLAIMS OF THIRD PARTIES, EVEN IF THE COMPANY OR AN AUTHORIZED REPRESENTATIVE OF THE COMPANY HAS BEEN ADVISED OF THE POSSIBILITY OF SUCH DAMAGES. IN NO EVENT SHALL LIABILITY OF THE COMPANY FOR DAMAGES WITH RESPECT TO THE SOFTWARE EXCEED THE AMOUNTS ACTUALLY PAID BY YOU, IF ANY, FOR THE SOFTWARE. SOME JURISDICTIONS DO NOT ALLOW THE LIMITATION OF IMPLIED WARRANTIES OR LIABILITY FOR INCIDENTAL, INDIRECT, SPECIAL, OR CONSEQUENTIAL DAMAGES, SO THE ABOVE LIMITATIONS MAY NOT ALWAYS APPLY. THE WARRANTIES IN THIS AGREEMENT GIVE YOU SPECIFIC LEGAL RIGHTS AND YOU MAY ALSO HAVE OTHER RIGHTS WHICH VARY IN ACCORDANCE WITH LOCAL LAW.

## ACKNOWLEDGMENT

YOU ACKNOWLEDGE THAT YOU HAVE READ THIS AGREEMENT, UNDERSTAND IT, AND AGREE TO BE BOUND BY ITS TERMS AND CONDITIONS. YOU ALSO AGREE THAT THIS AGREEMENT IS THE COMPLETE AND EXCLUSIVE STATEMENT OF THE AGREEMENT BETWEEN YOU AND THE COMPANY AND SUPERSEDES ALL PROPOSALS OR PRIOR AGREEMENTS, ORAL, OR WRITTEN, AND ANY OTHER COMMUNICATIONS BETWEEN YOU AND THE COMPANY OR ANY REPRESENTATIVE OF THE COMPANY RELATING TO THE SUBJECT MATTER OF THIS AGREEMENT.

Should you have any questions concerning this Agreement or if you wish to contact the Company for any reason, please contact in writing at the address below.

Robin Short
Prentice Hall PTR
One Lake Street
Upper Saddle River, New Jersey 07458